Enhancing Science Education

This book helps meet an urgent need for theorized, accessible and discipline-sensitive publications to assist science, technology, engineering and mathematics educators. The book introduces Legitimation Code Theory (LCT) and demonstrates how it can be used to improve teaching and learning in tertiary courses across the sciences. LCT provides a suite of tools which sci-ence educators can employ in order to help their students grasp difficult and dense concepts. The chapters cover a broad range of subjects, including biology, physics, chemistry and mathematics, as well as different curriculum, pedagogy and assessment practices. This is a crucial resource for any science educator who wants to better understand and improve their teaching.

Margaret A.L. Blackie is Senior Lecturer, Department of Chemistry and Polymer Science, Stellenbosch University, South Africa.

Hanelie Adendorff is Senior Advisor, Faculty of Science, Centre for Teaching and Learning, Stellenbosch University, South Africa.

Marnel Mouton is Senior Lecturer, Department of Botany and Zoology, Stellenbosch University, South Africa.

Legitimation Code Theory
Knowledge-Building in Research and Practice
Series editor: Karl Maton
LCT Centre for Knowledge-Building

This series focuses on Legitimation Code Theory or 'LCT', a cutting-edge approach adopted by scholars and educators to understand and improve their practice. LCT reveals the otherwise hidden principles embodied by knowledge practices, their different forms and their effects. By making these 'legitimation codes' visible to be learned or changed, LCT work makes a real difference, from supporting social justice in education to improving design processes. Books in this series explore topics across the institutional and disciplinary maps of education, as well as other social fields, such as politics and law.

Other volumes in this series include:

Accessing Academic Discourse
Systemic Functional Linguistics and Legitimation Code Theory
Edited by J. R. Martin, Karl Maton and Y. J. Doran

Building Knowledge in Higher Education
Enhancing Teaching and Learning with Legitimation Code Theory
Edited by Chris Winberg, Sioux McKenna and Kirstin Wilmot

Teaching Science
Knowledge, Language, Pedagogy
Edited by Karl Maton, J. R. Martin and Y. J. Doran

Decolonizing Knowledge and Knowers
Struggles for University Transformation in South Africa
Edited by Mlamuli Nkosingphile Hlatshwayo, Hanelie Adendorff, Margaret A.L. Blackie, Aslam Fataar and Paul Maluleka

Enhancing Science Education
Exploring Knowledge Practices with Legitimation Code Theory
Edited by Margaret A.L. Blackie, Hanelie Adendorff and Marnel Mouton

Sociology of Possibility
An Invitation to Legitimation Code Theory
Karl Maton

For a full list of titles in this series, please visit: https://www.routledge.com/Legitimation-Code-Theory/book-series/LMCT

Enhancing Science Education

Exploring Knowledge Practices with
Legitimation Code Theory

**Edited by
Margaret A.L. Blackie,
Hanelie Adendorff, and
Marnel Mouton**

Routledge
Taylor & Francis Group

LONDON AND NEW YORK

Cover image: © Getty Images

First published 2023
by Routledge
4 Park Square, Milton Park, Abingdon, Oxon OX14 4RN

and by Routledge
605 Third Avenue, New York, NY 10158

Routledge is an imprint of the Taylor & Francis Group, an informa
business

British Library Cataloguing-in-Publication Data
A catalogue record for this book is available from the British Library

Library of Congress Cataloging-in-Publication Data
A catalog record has been requested for this book

ISBN: 978-0-367-51869-1 (hbk)
ISBN: 978-0-367-51870-7 (pbk)
ISBN: 978-1-003-05554-9 (ebk)

DOI: 10.4324/9781003055549

Typeset in Galliard
by SPi Technologies India Pvt Ltd (Straive)

Contents

List of figures   *vii*
List of tables   *x*
List of contributors   *xi*

1 Enacting Legitimation Code Theory in science education   1
MARGARET A.L. BLACKIE, HANELIE ADENDORFF, AND MARNEL MOUTON

PART I
Academic Support in Science   19

2 Becoming active and independent science learners:
Using autonomy pathways to provide structured support   21
KAREN ELLERY

PART II
Physical Sciences   39

3 Improving assessments in introductory Physics courses:
Diving into Semantics   41
CHRISTINE M. STEENKAMP AND ILSE ROOTMAN-LE GRANGE

4 Building complexity in Chemistry through images   63
ZHIGANG YU, KARL MATON, AND YAEGAN DORAN

5 Using variation in classroom discourse: Making
Chemistry more accessible   82
BRUNO FERREIRA DOS SANTOS, ADEMIR DE JESUS SILVA JÚNIOR, AND
EDUARDO FLEURY MORTIMER

6 Radiation Physics in theory and practice: Using
Specialization to understand 'threshold concepts'   103
LIZEL HUDSON, PENELOPE ENGEL-HILLS, AND CHRIS WINBERG

PART III
Biological Sciences   **127**

7 Interdisciplinarity requires careful stewardship of powerful
knowledge   129
GABI DE BIE AND SIOUX MCKENNA

8 Advancing students' scientific discourse through
collaborative pedagogy   148
MARNEL MOUTON, ILSE ROOTMAN-LE GRANGE, AND BERNHARDINE UYS

9 Using Autonomy to understand active teaching
methods in undergraduate science classes   169
M. FAADIEL ESSOP AND HANELIE ADENDORFF

PART IV
Mathematical Sciences   **191**

10 A conceptual tool for understanding the complexities
of mathematical proficiency   193
INGRID REWITZKY

11 Supporting the transition from first to second-year
Mathematics using Legitimation Code Theory   206
HONJISWA CONANA, DEON SOLOMONS, AND DELIA MARSHALL

PART V
Science Education Research   **225**

12 Navigating from science into education research   227
MARGARET A.L. BLACKIE

Index   241

Figures

1.1 The specialization plane   6
1.2 The epistemic plane   7
1.3 The social plane   8
1.4 The semantic plane   10
1.5 The autonomy plane   13
2.1 The autonomy plane   24
2.2 Autonomy pathways for self-reflection intervention   27
2.3 Autonomy pathways for modelling consolidation of a lecture   29
2.4 Autonomy pathways for guided reflection on a test   31
3.1 The semantic plane   43
3.2 Semantic density maps of the Ph101 assessments of 2016 and
     2017 representing the percentage of marks allocated to
     (a) questions and (b) model answers of different semantic
     density categories   51
3.3 Number of question-answer pairs plotted on a two-dimensional
     graph representing the semantic density of answers versus
     questions, for the main tests of 2016 (a) and 2017
     (b), respectively   52
3.4 The questions of the class tests of 2016 (a) and 2017 (c)
     and first exams of 2016 (b) and 2017 (d) of Ph101,
     plotted on the semantic plane   53
3.5 The model answers of the class tests of 2016 (a) and 2017 (c)
     and first exams of 2016 (b) and 2017 (d) of Ph101, plotted
     on the semantic plane   54
3.6 The average marks plotted versus the semantic density of
     the question (a) and answer (c) and the semantic gravity (e),
     respectively, shown with the weights of the different categories
     of questions (b) and answers (d) plotted on the semantic plane   56
3.7 Average marks in the first exams of 2016 and 2017 as a
     function of the SDD   59
4.1 Melting ice cubes (top); the electrode reactions in a
     Daniell cell (bottom)   64
4.2 Fresh water in a cup (left); hydrogen bonding between
     water molecules (right)   67

4.3	Compressibility of air (left); a structural formula of water molecules (right)	68
4.4	A structural formula of methane molecules	69
4.5	Two pathways for the production of ibuprofen	70
4.6	A pile of matches (left); a burning match (right)	71
4.7	A campfire (left); energy change during combustion of carbon (middle); an energy level diagram for the formation of carbon dioxide from carbon and oxygen carbon monoxide (right)	72
4.8	A rusted car (left); a diagram of a rusty nail experimental set-up (right)	73
4.9	Two segments of orange (left); the formation of sodium chloride	74
4.10	Batteries (left); a diagram showing the formation of a secondary amide through a condensation reaction between ethanoic acid and methylamine (right)	75
4.11	A widening range of images' epistemic–semantic density across stages	76
4.12	Ice is less dense than liquid water	77
4.13	The addition reaction of ethene with bromine	77
4.14	The formation of the hydronium ion	78
5.1	The triplet relationship	83
5.2	An upward shift or 'packing' movement in the classroom discourse	88
5.3	A downward shift	91
5.4	An upward semantic wave shift	92
5.5	A 'downward escalator' profile	93
5.6	A 'flatline' profile	94
5.7	A downward shift followed by an upward shift	96
6.1	A relational view of the features of threshold concepts	105
6.2	The specialization plane	107
6.3	The specialization plane for radiation physics	108
6.4	Plotting the threshold concept framework on the specialization plane	120
7.1	The specialization plane	132
7.2	Three semantic profiles	134
7.3	Specialization codes of the two professions	144
8.1	Wording tool for epistemic–semantic density in English discourse	155
8.2	Bar graph indicating proportion of simpler to more complex meaning in the respective descriptions of the nucleus of a eukaryotic cell from the first-year and the school textbooks	158
8.3	Bar graph indicating proportion of simpler to more complex meaning in the respective descriptions of the nucleus of a eukaryotic cell from the students' (1 to 6) final summative assessments	159

8.4	A–D Concept maps of students' biology vocabulary describing the nucleus of a eukaryotic cell	160
8.5	Concept map of student 6's biology vocabulary describing the nucleus of a eukaryotic cell	161
8.6	Semantic density profile of concept 1 component 2 from first-year and school textbooks, and from final summative assessment of student 4	164
9.1	The autonomy plane	176
9.2	Autonomy plane for active teaching methods in physiology	177
9.3	Example of an autonomy tour	178
9.4	The autonomy tour traced by the first four task steps in a Can of Bull	182
9.5	Example of a Burning Question	183
9.6	The return trip traced by the steps in the Burning Question	183
9.7	Running Question – first information slide	185
9.8	Running Question – second information slide	185
9.9	Running Question – theory slide	186
10.1	The specialization plane	196
10.2	The epistemic plane	198
10.3	Insights for functions	200
10.4	Insights for a mathematical activity	201
10.5	Insights for linking mathematical representations of a mathematical object	202
10.6	Insights for the concept of mathematical structure	203
11.1	Illustrative profiles and semantic ranges	208
11.2	The specialization plane	210
11.3	Semantic profile of a first-year ECP physics lesson	213
11.4	Semantic profile of a second-year physics lesson	213
11.5	Semantic profile of a second-year mathematics lesson	214
12.1	Nested relationship between the real, the actual and the empirical domains	228

Tables

3.1 Translation device for semantic density   47
3.2 Examples of the coding of assessment questions representing different categories of semantic density and semantic gravity, along with an explanation to justify the allocated coding   48
3.3 Translation device used for semantic gravity levels   50
4.1 A translation device for the epistemic–semantic density of images in secondary school chemistry textbooks   67
5.1 Translation device for epistemic–semantic density of chemistry knowledge   86
5.2 Translation device for epistemic–semantic density of words   88
5.3 Word-grouping tool for epistemic–semantic density in English discourse   88
5.4 Highest and lowest strength of epistemic–semantic density in Carlos's and Bento's episodes   91
5.5 Stronger and weaker strength of epistemic–semantic density in Durval's and Marina's episodes   94
5.6 Stronger and weaker strength of epistemic–semantic density in Carlos's and Marina's episodes   96
6.1 Using specialization codes to understand threshold concepts   121
8.1 Epistemic–semantic density categories of Maton and Doran (2016) with descriptions and examples from students' discourse   154
8.2 An adjusted version of the word-grouping tool for epistemic–semantic density   156
9.1 Positional autonomy – a translation device   179
9.2 Relational autonomy – a translation device   179
9.3 Running Question 'autonomy tour' with multiple return trips   187
10.1 Ontic relations and discursive relations for a particular mathematical object of study   198
11.1 Translation device for various levels of Semantics   212

Contributors

Hanelie Adendorff is a Senior Advisor in the CTL at Stellenbosch University, South Africa. She has a PhD in chemistry but has been working in professional development since 2002. Her career and professional development started with an interest in blended learning and has since included the areas of assessment, facilitation of collaborative learning, science education and, more recently, decolonization of the science curriculum. As a member of the Faculty of Science's teaching and learning hub, she works with the Vice-Dean (Teaching and Learning) to enhance the status of teaching in the Faculty.

Margaret A.L. Blackie is a Senior Lecturer in the Department of Chemistry and Polymer Science at Stellenbosch University, South Africa. Since starting her independent career, she has attempted to hold together research interests in synthetic organic chemistry and in education research. She is passionate about understanding how knowledge in chemistry is built in the field and in the minds of students. She has been the recipient of several awards, including the South African Chemical Institute Chemistry Education Medal and the Stellenbosch University Distinguished Teacher Award. She is also the recipient of the Stellenbosch University Teaching Fellowship focusing on creating resources to smooth the path for academics from the natural sciences to engage with education research.

Honjiswa Conana is a Teaching and Learning Specialist in the Faculty of Natural Sciences at the University of the Western Cape, South Africa. Her role is to support teaching and learning initiatives in the Faculty. She works alongside lecturers in helping them to make explicit to students the aspects of the disciplinary discourse. Prior to this appointment, she worked as an academic literacy practitioner in the Department of Physics and Astronomy, working in collaboration with first- and second-year Physics lecturers. In this role, Dr Conana would help lecturers to identify the disciplinary practices, as well as assist them in developing classroom activities to make these practices explicit to students. Her research interests lie in physics education and academic literacies. Her research has been focused on studying students' experiences in a curriculum that is explicitly designed to develop physics students' academic literacy and their access to the disciplinary discourse of physics.

Gabi de Bie was a passionate and committed educator who served as chief scientific officer in the Department of Human Biology at the University of Cape Town, South Africa from 2009 to 2017. She was always as concerned about her students' social well-being as she was about their intellectual development. She was also passionate about environmental sustainability and social justice. She undertook her PhD at Rhodes University and graduated in 2016. She died of cancer in October 2020 and is very much missed by family, friends and colleagues.

Ademir de Jesus Silva Júnior is Professor of Chemistry Education at the Department of Exact and Natural Sciences, State University of Southwestern Bahia, Brazil. His research interests include Chemistry education, classroom discourse and students' concept building.

Y. J. Doran is a Postdoctoral Researcher at the University of Sydney, Australia, who focuses on language, semiosis, knowledge and education. His research spans the interdisciplinary fields of educational linguistics, discourse analysis, language description and multimodality and focuses mainly on English and Sundanese.

Karen Ellery is an Associate Professor in the Centre for Higher Education Research, Teaching and Learning at Rhodes University, South Africa, where her primary function is to coordinate and teach in the Science Extended Studies Programme. Her research centres on enabling epistemological access in the sciences in a higher education context, focusing specifically on curriculum structures, pedagogic practices and student learning.

Penelope Engel-Hills retired as Associate Professor: Interdisciplinary Health Sciences and Dean: Faculty of Health and Wellness Sciences at the Cape Peninsula University of Technology (CPUT) in January 2021. She is now an Adjunct Professor at CPUT with responsibilities in the faculty and Professional Education Research Institute (PERI). Prof Engel-Hillis has intersecting roles as a health care practitioner, educator and researcher. She currently holds ministerial appointments on the councils of the National Health Research Ethics Council, the Health Professions Council of South Africa and a Technical and Vocational Education and Training College. She has graduated and continues to supervise a number of postgraduate students and co-hosts a postdoctoral fellow in an interdisciplinary research role at the Applied Microbial and Health Biotechnology Institute.

M. Faadiel Essop is currently a Professor in the Department of Physiological Sciences at Stellenbosch University, South Africa, and the Director of the recently established Centre for Cardio-metabolic Research Centre in

Africa (CARMA) at Stellenbosch University. Prof Essop is an NRF-rated internationally acclaimed researcher, former chairperson of the Department of Physiological Sciences, a former President of the Physiology Society of Southern Africa, the current Vice-President of the African Association of Physiological Sciences and a board member of the General Assembly of the International Union of Physiological Sciences. He also served as a member of the International Committee of the American Physiological Society and is an elected Fellow of the American Physiological Society.

Bruno Ferreira dos Santos is a Professor of Chemistry Education at the Department of Sciences and Technologies at the State University of Southwestern Bahia, Brazil. He works with chemistry teacher education and in-service education for science teachers. His research interests include teaching and learning sciences, disciplinary learning and its relationship to language and other semiotic resources.

Eduardo Fleury Mortimer is a Professor of Education at the Federal University of Minas Gerais, Brazil and a researcher at the National Council for Scientific and Technological Development. His research interests focus on science learning, conceptual profiles and classroom discourse. He is the former president of the Brazilian Association for Research in Science Education (2005–9).

Lizel Hudson is the Work-Integrated Learning and Language Coordinator in the Faculty of Health and Wellness Sciences at the CPUT. Her research interests include health professions education and in particular knowledge-building in simulated learning environments. She recently completed her doctoral studies, which focused on understanding threshold concepts in professional education. Lizel is currently participating in the Teaching Advancements at University Fellowships Programme where her project focuses on smart student engagement and a holistic experience towards student success.

Delia Marshall is a Professor in the Department of Physics and Astronomy at the University of the Western Cape (UWC), South Africa, where she is currently Head of Department. Her areas of research include undergraduate physics education and higher education studies. She has a particular interest in students' experiences of the transition from high school to university studies and has been involved in the UWC Science Faculty's extended curriculum programme. Working with Honjiswa Conana, her research on physics learning has used Legitimation Code Theory (LCT) to examine how students take on the disciplinary discourse of physics, as well as students' use of representations in problem-solving. Her current research interests include the role of higher education in society and for the public good, the notion of 'graduateness' in the context of the undergraduate education of physics graduates and the post-graduation trajectories of young South Africans.

Karl Maton is Professor of Sociology at the University of Sydney, Australia; Director of the LCT Centre for Knowledge-Building; Visiting Professor at the University of the Witwatersrand, South Africa; and Visiting Professor at Rhodes University, South Africa. His book *Knowledge and Knowers: Towards a Realist Sociology of Education* is the founding text of LCT (2014). He has also recently co-edited *Knowledge-Building: Educational Studies in Legitimation Code Theory* (2016), *Accessing Academic Discourse: Systemic Functional Linguistics and Legitimation Code Theory* (2020) and *Teaching Science: Knowledge, Language, Pedagogy* (2021). He is currently writing *Legitimation Code Theory: A Primer.*

Sioux McKenna is a Professor of Higher Education and is director of the Centre for Postgraduate Studies at Rhodes University, South Africa. Her research reflects a concern about who gets access to knowledge in the academy and whose knowledge is valued. She is also interested in how the norms and values of disciplines emerge as literacy practices. She is the project manager of a number of international research projects and has supervised a number of postgraduate studies, many of which have used LCT to ask questions about the form and function of higher education.

Marnel Mouton is a Senior Lecturer in the Department of Botany and Zoology at Stellenbosch University, South Africa. She has a PhD in Microbiology and currently teaches Biology to various cohorts of under-graduate students studying programmes in Science, AgriScience and Health Science. She is also a lecturer and the coordinator of the Extended Degree Programme for the Faculties of Science, AgriScience and Engineering at Stellenbosch University. These roles led to her keen inter-est and research projects in evaluating and addressing the articulation gap between school and first year in higher education and specifically the life sciences.

Ingrid Rewitzky is involved in educational leadership at Stellenbosch University, South Africa, within and across relatively complex environ-ments through her roles as Professor of Mathematics, Executive Head of the Department of Mathematical Sciences and Vice-Dean (Learning and Teaching) of the Faculty of Science and through serving on committees in the department, faculty and university. As Vice-Dean, she has been instrumental in facilitating and supporting a scholarly approach to teach-ing and learning in Science at Stellenbosch University. In December 2018, she was honoured with the Stellenbosch University Chancellor's Award in recognition of sustained excellence throughout her career. Her personal growth and satisfaction as an academic and educational leader are integrally linked to understanding and establishing connections between a diversity of perspectives.

Ilse Rootman-le Grange is the E-learning Instructional Designer for the Faculty of Science at Stellenbosch University, South Africa. She supports lecturers in Science in the development of teaching practices that include learning technologies. She has a PhD and teaching experience in Chemistry. Her research interests in undergraduate science education include modes of teaching, bridging the articulation gap, the role of multidisciplinary collaborations in science education and professional development of undergraduate lecturers in science.

Deon Solomons is a Senior Lecturer in the Department of Mathematics and Applied Mathematics at the University of the Western Cape (UWC), South Africa. He teaches real analysis and modern algebra to third-year students, measure and integration theory and computational linear algebra at honours level. He provides postgraduate supervision in differential geometry and gravitation, and in mathematical physics. He conducts research in these two fields in addition to mathematics education. Dr Solomons' also helps out with the UWC MathClub and media and marketing for the Department of Mathematics and Applied Mathematics.

Christine M. Steenkamp is a Senior Lecturer in Physics at Stellenbosch University, South Africa, and holds a PhD in laser physics. Her interest in LCT is aimed at improving undergraduate teaching in physics, with a focus on the first-year physics modules. She is involved in undergraduate teaching and postgraduate supervision. Her current research projects in physics involve laser spectroscopy, quantum systems and methods to purify isotopes for medical purposes.

Bernhardine Uys is based in Stellenbosch University, South Africa, as a part-time Research Assistant. She has worked with Dr Marnel Mouton on projects pertaining to LCT and its application in the field of Science education to bridge the articulation gap and increase the ability of students to develop skills in scientific understanding and argumentation. Her interests lie in Education and Journalism, having acquired a B.Ed. and a B.Hons in Journalism from Stellenbosch University and having subsequently worked in both fields over the years.

Chris Winberg holds the South African National Research Foundation Chair in Work-Integrated Learning and leads the PERI at the CPUT in Cape Town, South Africa. Chris currently serves as a member of the South African Ministerial Oversight Committee on Transformation in Higher Education. Her research focus is professional and vocational education – with a particular focus on engineering education, the professional development of university teachers and technical communication. She obtained a PhD in applied linguistics from the University of Cape Town and lectured in applied linguistics and language education at the University of Cape Town, the University of the Western Cape and the University of Stockholm in Sweden. From 2010 to 2012, she was chairperson of the South African Association for Applied Linguistics. Chris was director of the Fundani Centre for Higher Education Development at the CPUT 2011–15, responsible for supporting curriculum renewal, academic staff development and promoting educational research.

Zhigang Yu is a Lecturer at Beijing Institute of Technology, China, who focuses on semiotics and education using LCT and systemic functional linguistics. He undertook his PhD at Tongji University, China, and studied at the University of Sydney under the supervision of Karl Maton and Y. J. Doran. His current research interests include multimodality in chemistry discourse and knowledge-building of secondary school chemistry.

1 Enacting Legitimation Code Theory in science education

Margaret A.L. Blackie, Hanelie Adendorff, and Marnel Mouton

Introduction

The purpose of science education is not only to produce a new generation of scientists but also to offer young people understanding about the vast explanatory power that science has to offer to a world faced with complex challenges. Debates on how to handle existential threats such as global warming and diseases like COVID-19 are good examples of such problems. Osborne and Dillon (2008) argue that science education 'should be to educate students both about the major explanations of the material world that science offers and about the way science works.' Moreover, scientific understanding and reasoning are desired attributes for the future citizen in many countries of the world. Wide-ranging studies have shown that this aspiration will require dedicated investments in skilled science educators that continuously develop their own knowledge and teaching practice, the development of genuinely engaging curricula, as well as assessment protocols and structures that will meet the desired outcomes and goals. Currently, we are not achieving this ideal. Many European countries, for example, have witnessed a decline in the number of students who enrol for degrees in science (Osborne and Dillon 2008). Moreover, science education continues to represent a substantial hurdle for both lecturers and students around the globe. Students find science courses difficult to master, and lecturers find science courses challenging to teach, although the nature of that challenge varies from subject to subject (Sithole *et al.* 2017). Nonetheless, science education has tended to focus on the mastery of particular scientific concepts rather than on the induction of the student to the knowledge field as a whole.

Many academic scientists are interested in developing and improving their own pedagogy but may struggle to find a 'way into' engaging with the scholarship of teaching and learning (Adendorff 2011). Conducting science education research is an even bigger challenge. The stumbling block for many academics making this transition is the apparent lack of clarity on the links between methods and theoretical frameworks (Adendorff 2011). Whilst some educationalists have attributed the lack of impact of their work on the practices of many science educators to an arrogant dismissal of education as unscientific, this is far too simplistic and one-sided. Differences of

DOI: 10.4324/9781003055549-1

terminology, methodology, style and even epistemology can make research into science education appear daunting and alien to university-based scientists. Moreover, some approaches in science education research are less than convincing with the use of vague terminology, loose logic and minimal empirical evidence. There is thus a need for an approach that is clear, explicit, evidential and rigorous, to help engage scientists with scholarship that can enhance their pedagogic practices and enable science education research. This book brings together a rich collection of studies in science education that uses a common framework, Legitimation Code Theory, to attend to these concerns.

Legitimation Code Theory (LCT) provides a welcome entry point into a scholarly approach to science pedagogy, as well as rigorous science education research. Moreover, we have found that academic scientists experience LCT as more 'science-like' and therefore 'less foreign,' relative to other education research frameworks. LCT offers a suite of tools which can be used for a wide range of purposes (which is explained later in this chapter). For example, it may be used to analyze conceptual gain, or to evaluate the ways in which knowledge and social relations interact in a particular situation or to interrogate the aims and purposes of different learning activities in a course. We can use the framework to examine from the ways in which scientific concepts are taught to the ways we structure science assessments (Rootman-le Grange and Blackie 2018, 2020; Steenkamp *et al.* 2019).

There is already a diversity of ways in which LCT has been enacted for evaluating and shaping science pedagogy and curricula, with the framework finding application in Biology (Mouton and Archer 2019; Mouton 2020), Chemistry (Blackie 2014) and Physics (Georgiou 2016), to name but a few. LCT comprises several 'dimensions' or sets of concepts, one of which is Semantics. The following three examples all make use of Semantics in different ways. Conana *et al.* (2016) analyzed the way in which language and concepts were used in an introductory Physics course. These authors showed that the lecturer almost exclusively used specialist language, which was troublesome for students to access. LCT holds that knowledge-building for epistemological access requires waves of movement between simpler, everyday language and the more specialist language of the subject (Maton 2009). Kelly-Laubscher and Luckett (2016) used Semantics to show that there is a vast disparity in complexity between high school and university Biology textbooks. This difference may be one of the reasons why students who achieved good marks in school struggled with the subject in their first year at university. Mouton and Archer (2019) and Mouton (2020) have since built on these findings in Biology to develop a pedagogy and learning activities to mitigate the articulation gap between school and first year in higher education. In all of these studies, LCT was used to reveal tacit problems and to shape teaching practice to overcome the problems. These are just some possibilities among many. LCT offers the possibility of a breadth of exploration at any level – from a single lecture or practical, to an entire degree programme.

Through the exploration of LCT, two further aspects of education have come into view: cumulative knowledge-building (Maton 2009) or extending existing ideas and integrative knowledge-building (Maton and Howard 2018) or productively bringing together different ideas. Science students often struggle to recognize particular scientific concepts in a different context, a key outcome of most science programmes. Cumulative knowledge-building is essential to ensure that a student will be able to use concepts and language beyond the scope of the particular course, such as the capacity to use science concepts of acids and bases taught in an introductory Chemistry course in a second-year Biochemistry course or in real-world problems pertaining to acid rain. Integrative knowledge-building is the integration of different kinds of knowledge – this is the foundation for lifelong learning and key to solving real-life, complex problems. For example, recognizing that developing new, healthier and cheaper or more sustainable food products draws from knowledge in Chemistry, Biochemistry and Microbiology. Similarly, understanding the mechanism of infection of a virus such as SARS-CoV-2, the cause of COVID-19, requires drawing knowledge from different disciplines.

The methodology of LCT also offers a significant advantage to academics who have been used to disciplinary STEM-based research. Feedback from various workshops on LCT suggests that academic scientists find that the integral use of Cartesian planes (described later in this chapter) offers a familiar visual framework which somehow makes LCT feel more 'science friendly.' The take up of LCT in the science community speaks for itself.[1] This emerging body of work shows how academics across scientific disciplines have used LCT in the analysis and shaping of their current teaching practice. To date, such efforts have been largely *ad hoc*. The vast majority of papers have come from a relatively small community of science educators who have stumbled across LCT and found it very useful, though the rapid growth of this community in recent years is reaching a critical mass of productivity. It is thus timely to gather a collection of these efforts to show something of the range of what the use of LCT can achieve within a scientific context.

In this chapter, we introduce the conceptual framework, LCT, to science educators. We look at each of the LCT dimensions that are used throughout this book – Specialization, Semantics and Autonomy. At the end of this chapter, we present a brief summary of the chapters that reach across the sciences and which embrace curriculum design, pedagogic practice and assessment.

Legitimation Code Theory

LCT is a realist framework, developed by Karl Maton (2014), which builds on the work of Basil Bernstein and Pierre Bourdieu, among others. It offers a multi-dimensional approach for exploring what it means to know and how one comes to know in different disciplines or knowledge practices (Winberg *et al.* 2020). The sociologists Bernstein and Bourdieu both witnessed the wave of massification of higher education, which shifted the demographic of the student body from a small, privileged élite to a large diverse group including more social classes. It soon became apparent that this greater access did not translate to success for all since not all students had the cultural and social capital required to engage meaningfully in higher education. This has been highlighted by Morrow's (2009) work on 'epistemological access' to the required knowledge. The work of Bernstein (2000), Bourdieu (1988) and Morrow (2009) aim to expose some of the impediments to entry into academia. LCT has a similar social justice agenda – making the 'rules of the game' explicit to all participants, potentially affording access to those who have not been culturally conditioned to see the dynamics in play (Maton 2014).

One of the ways in which LCT does this is by addressing the issue of *knowledge-blindness*, where knowledge is reduced to knowing (mental processes of understanding) whilst losing sight of the organizing principles at play in different knowledge practices (Maton, 2014: 3). With its focus on revealing these underlying logics, LCT allows us to show the ways in which coming to 'know' differ across different knowledge practices. LCT's set of tools can be enacted to explore knowledge, i.e. what counts as a legitimate claim, who is allowed to make such a claim, and how meaning is made by making explicit that which is often hidden or tacit and taken for granted. Its various concepts and dimensions offer a means to reveal different aspects of these 'rules of the game' in diverse practices.

Dimensions of LCT

Three of LCT's dimensions are well developed and in fairly wide usage: Specialization, Semantics and Autonomy. Specialization and Semantics are both thoroughly described in *Knowledge and Knowers* (Maton 2014) and in *Knowledge-Building* (Maton *et al.* 2016). Autonomy was not fully developed at that time, but an extensive overview of this dimension was presented in a paper written by Maton and Howard (2018). As mentioned earlier, these three dimensions allow exploration of different aspects of knowledge practices. Specialization is focused on how knowledge and knowers are legitimated in different knowledge practices. Semantics reveals how meaning is made. Autonomy explores the origin and purposes of various constituents of knowledge practices. Each LCT dimension is conceived as a combination of two organizing principles. These two organizing principles are independent of one another, each with the ability to vary from weaker to stronger, and can thus be plotted on a Cartesian plane with each of the principles represented by one of the axes. Practices can valorize one, both or neither of the organizing principles, leading to four overarching modalities for each dimension, which are called 'codes.'

Specialization

Specialization focuses on the basis for legitimacy in different practices, i.e. who can make a legitimate knowledge claim, as well as what would constitute a legitimate knowledge claim. This starts from the perspective that all knowledge claims are about something and made by someone. The organizing principles in the case of Specialization are *epistemic relations* (ER), between the knowledge practice and its objects, and *social relations* (SR), between the practice and its subjects. Fields with relatively strong epistemic relations (ER+) place emphasis on knowledge, skills and procedures whilst fields with relatively strong social relations (SR+) valorize dispositions, values and attributes of knowers (Maton 2014). The two relations can be plotted as the *specialization plane*, with four principle modalities or *specialization codes* as shown in Figure 1.1 (Maton *et al.* 2016). Knowledge practices are always underpinned by epistemic relations and social relations, but it is the degree to which each organizing principle is emphasized that determines the basis of achievement in a particular practice. As stated above, practices can emphasize one, both or neither of these relations as a basis for legitimacy whilst both relations can vary from stronger to weaker, allowing an infinite number of strengths or positions on the specialization plane (Figure 1.1).

The principal modalities or *specialization codes* are (Figure 1.1):

- *knowledge codes* (ER+, SR−) arise when we have stronger epistemic relations (ER+) coupled with weaker social relations (SR−), i.e. where practices emphasize the possession of specialized skills, knowledge or procedures as the basis for success whilst downplaying the attributes of the actor making the claim. In this code, what one knows is important, and one's dispositions may be gently overlooked. Legitimate participation in the natural sciences is often dominated by different variations of this code.
- *élite codes* (ER+, SR+) arise when stronger epistemic relations (ER+) are coupled with stronger social relations (SR+), i.e. where practices emphasize the possession of both specialized skills, knowledge or procedures and attributes of the actor making the claim. In this code, both what one knows and who you are provide the basis for legitimacy. Fields that are both technically demanding and require some kind of individual expression, such as professional classical music performance, may be dominated by this code.
- *knower codes* (ER−, SR+) arise when weaker epistemic relations (ER−) are coupled with stronger social relations (SR+), i.e. where practices emphasize the attributes of the actor making the claim and downplay the possession of specialized skills, knowledge or procedures as the basis for legitimacy. In this code, who one is, is important, not what one knows. Many practices in the humanities are dominated by this code, through notions of a cultivated gaze.
- *relativist codes* (ER−, SR−) arise when legitimacy is determined by neither one's specialist knowledge nor one's personal attributes. This is a sort of 'anything goes,' such as when brainstorming without limits on what is a permissible idea to add.

Knowledge of the dominant code in a practice can help us unpack the rules for legitimacy, or the basis for achievement, in that practice. Besides the

knowledge   *élite*

SR–   SR+

social
relations

relativist   *knower*

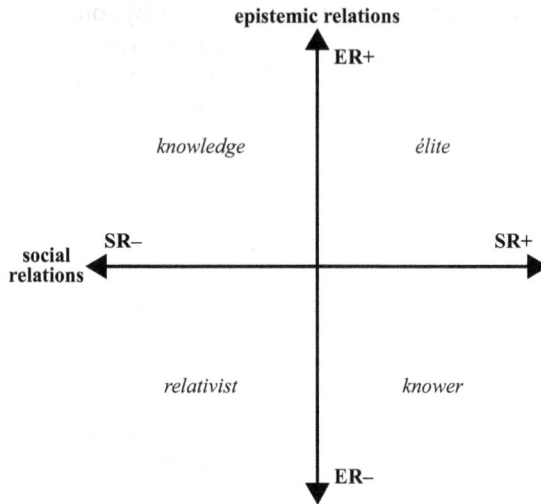

Figure 1.1 The specialization plane (Maton 2014: 30).

ability to change or shift over time, codes can also match, i.e. when two sets of practices use the same basis for success, or codes can clash. Code clashes occur when people or practices are characterized by different codes. Scientists who undertake education research for the first time often experience such a *code clash* when introduced to literature in teaching and learning that uses a knower code as its basis for claims. This might also be one reason why LCT, with its stronger epistemic relations, has found traction in many science environments.

Each of these organizing principles – epistemic relations and social relations – can be explored in more detail. 'Epistemic relations' can be broken down into 'what practices relate to and how they so relate' (Maton 2014: 174). These are *ontic relations* (OR) between knowledge practices and their objects of study, and *discursive relations* (DR) between knowledge practices and other knowledge practices (such as between different theories and methods). These relations can be plotted on the *epistemic plane* (see Figure 1.2), allowing us to distinguish four principal modalities or *insights*:

- *doctrinal insight* (OR–, DR+): what counts as legitimate objects of study is not tightly controlled (weaker ontic relations), but there are strong boundaries between what qualifies as a legitimate approach and what does not (stronger discursive relations). Legitimacy is thus the result of using a specialized approach.
- *situational insight* (OR+, DR–): strongly bounds and controls what can be legitimately studied (stronger ontic relations) but weakly bounds how this can be done (weaker discursive relations). What is studied is

significant for legitimacy, but there is relative flexibility in terms of approaches used.

- *purist insight* (OR+, DR+): both legitimate objects of study and legitimate approaches are strongly bound and thus significant.
- *knower/no insight* (OR–, DR–): both the objects of study and the legitimate approaches are weakly bound. Thus, neither the object of study nor the method of study is used as a basis for legitimacy. This may be *knower insight* when these weaker epistemic relations are paired with stronger social relations (a knower code or ER–, SR+), or it may be *no insight* when paired with weaker social relations (a relativist code or ER–, SR–).

The *epistemic plane* is useful for distinguishing between the kinds of knowledge that are being developed (Maton 2014). One of the major complaints of employers of science graduates is that they are unable to apply their knowledge. Among many possible applications, the epistemic plane can be used to explain why this might be. Lecturers may focus on the use of particular methods but fail to clearly show the limits of their application. This means that students may be able to pass courses and apply specific approved methods (DR+) to solve problems carefully chosen by the examiners (could be OR– or OR+). However, on entering employment, the new graduate is likely to be faced with complex or multifaceted problems and must then decide which methods can be legitimately applied. If the limits of application, i.e. variation in strength of OR, was not a major consideration in the course, the new graduate may struggle.

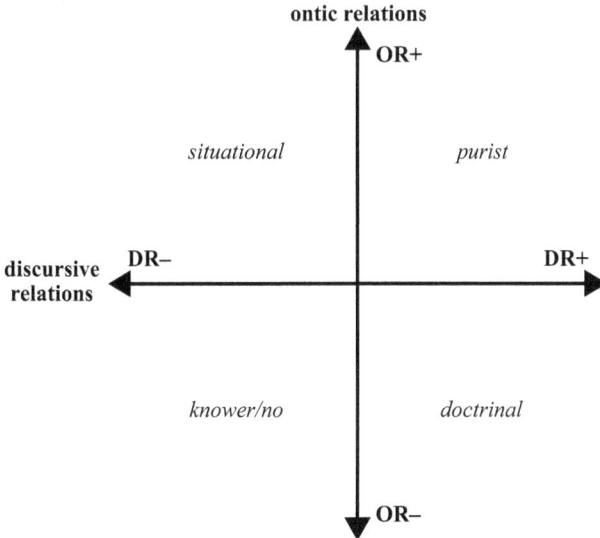

Figure 1.2 The epistemic plane (Maton 2014: 177).

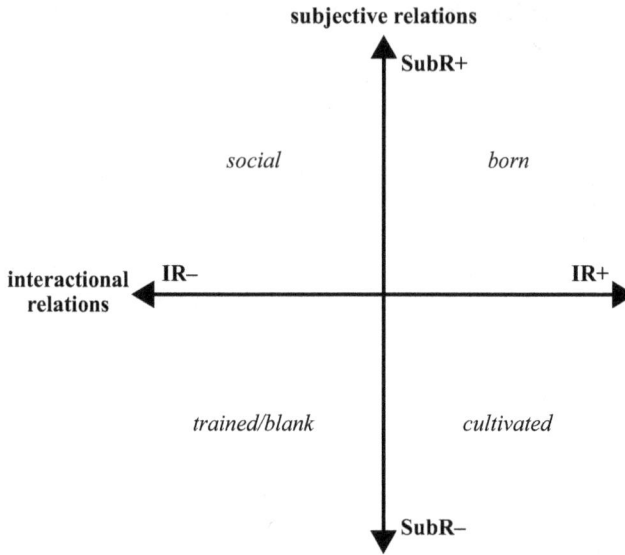

Figure 1.3 The social plane (Maton 2014: 186).

Turning to 'social relations,' the *social plane* (Figure 1.3) can be used to explore in greater depth different kinds of relations to knowers. These concepts are most applicable to knower-code practices (ER–, SR+) or élite-code practices (ER+, SR+), i.e. where social relations are relatively strong (Maton 2014). The social plane does not feature in this book. However, it is described here for the purposes of a more rounded introduction to the suite of tools most widely enacted at present. In addition, it affords the possibility of making the practices of knower-code fields (such as many parts of education research) more understandable to scientists. It may also be useful in science disciplines with a strong professional development orientation where social relations are also explicitly valued.

The *social plane* (Figure 1.3) maps the distinction between legitimation of practice on the basis of emphasis on who one is (*subjective relations*) and ways of knowing through interactions with significant others (*interactional relations*). Both can take many forms; for example, subjective relations may highlight social class, sex, gender, race, ethnicity, sexuality, religion, etc., and interactional relations may highlight prolonged immersion in a canon of great works, spending time within a culture and so on. Both relations may differ in how emphasized they are as the basis of legitimacy. As before, plotting these relations as a plane result in four modalities or *gazes*:

- *social gazes* (SubR+, IR–) emphasize legitimacy as a legitimate knower based on who one is (stronger subjective relations) and downplay the significance of specific ways of knowing (weaker interactional relations). An example is offered by standpoint theories that allow only those with a particular identity, such as being LGBTQIA+, to claim legitimacy.

- *cultivated gazes* (SubR–, IR+) emphasize legitimacy not on the basis of one's identity (weaker subjective relations) but rather on the basis of how one interactionally comes to be a knower (stronger interactional relations). These often involve acquiring a 'feel' for practices through, for example, extended participation in 'communities of practice,' sustained exposure to exemplary models, such as great works of art, and prolonged apprenticeship under an acknowledged master.
- *born gazes* (SubR+, IR+) emphasize both legitimate kinds of knowers (stronger subjective relations) and legitimate ways of knowing (stronger interactional relations), such as claims to legitimacy based on both membership of a social category and experiences with significant others (e.g. standpoint theory that additionally requires mentoring by already-liberated knowers in consciousness-raising groups).
- *trained/blank gazes* (SubR–, IR–) emphasize neither kinds of knowers nor ways of knowing as the basis of legitimacy. As part of specialization codes, they emphasize either stronger epistemic relations (trained gaze) or nothing at all (blank gaze).

Semantics

The Semantics dimension of LCT considers the nature of meanings in terms of context and complexity. The organizing principles are *semantic gravity* and *semantic density* (Maton 2014).

Semantic gravity (SG) refers to the degree to which meaning relates to its context (Maton 2013, 2014; Maton *et al.* 2016). Semantic gravity can be stronger and weaker along a continuum of strengths. When the meaning is strongly tied to a context, semantic gravity is stronger (SG+); when meaning is weakly tied to a context, semantic gravity is weaker (SG–). In practice, semantic gravity can be strengthened by moving from more decontextualized meanings to more concrete, contextualized meanings and weakened by doing the opposite. In science teaching, for example, real-world applications of theoretical concepts can be employed to strengthen semantic gravity, and then returning to the theoretical concepts would weaken semantic gravity.

Semantic density (SD) refers to the complexity of meaning (Maton 2013, 2014; Maton *et al.* 2016). Semantic density can also be stronger or weaker along a continuum of strengths. Stronger semantic density (SD+) indicates more complex meanings; weaker semantic density (SD–) indicates less complex meanings. In practice, semantic density can be dynamized by moving (strengthening and weakening) between more complex, condensed meanings and simpler meanings. In science teaching for example, when a scientific term or concept is introduced or used, the meaning is often relatively complex or stronger semantic density; when the lecturer then unpacks and explains these meanings using simpler words and terms, they are expressing weaker semantic density; then they return to the concept; they are moving back to stronger semantic density.

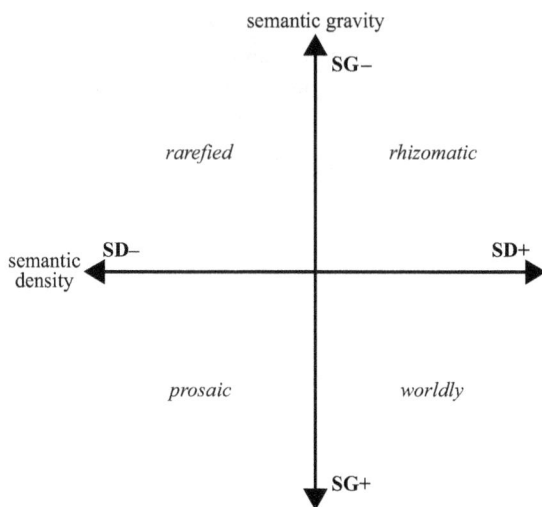

Figure 1.4 The semantic plane (Maton 2014: 131).

The strengths of the two organizing principles, *semantic gravity* and *semantic density*, may vary independently. These are mapped on the *semantic plane* (SG±, SD±): semantic gravity is the *y*-axis and semantic density is the *x*-axis, as shown in Figure 1.4 (Maton *et al.* 2016). We can identify four principal *semantic codes* (Maton 2013, 2014; Maton *et al.* 2016):

- *rhizomatic codes* (SG−, SD+), where meaning and the 'basis of achievement' is relatively context-independent (weaker semantic gravity) and complex and condensed (stronger semantic density). Examples in science education may include complex theoretical terms or abstract concepts, often expressed in specialist scientific language or symbols, where no external context is given or available.
- *worldly codes* (SG+, SD+), where legitimacy is based on meanings that are relatively context-dependent and more concrete (stronger semantic gravity) but complex and condensed (stronger semantic density). An example in science teaching may include teaching or using complex scientific terms or concepts taught against a backdrop of a real-world context.
- *prosaic codes* (SG+, SD−), where legitimacy represents meanings that are relatively context-dependent (stronger semantic gravity) and simpler (weaker semantic density). Examples of these codes in science teaching may include using simpler meaning (possibly everyday concepts or basic scientific terms) that apply to real-world contexts – maybe as a way to explain more complex content later in a lecture.
- *rarefied codes* (SG−, SD−), where legitimacy is based on meanings that are more context-independent (weaker semantic gravity) but relatively simple (weaker semantic density). Here, examples in science teaching may include the use of simpler theoretical terms, but without the

background of context (decontextualized), possibly purely theoretical, but relatively simpler meaning.

Practice (such as classroom practice) can and should ideally display code shifts on the semantic plane – movements between decontextualized and more contextualized meanings, as well as between simpler and more complex meanings. This shifting between semantic codes is known as *semantic waves* (Maton 2013, 2014). For example, Mouton and Archer (2019) have shown how pedagogy in Biology should enact semantic waves to facilitate cumulative learning and Mouton (2020) further showed how project-based learning can be employed to reach the same goal. Similarly, Blackie (2014) argues that many lecturers (organic Chemistry in her case) use terms and simply presume that students understand the broader scope of what is being said. Instead, lecturers should consciously and intentionally move between stronger and weaker semantic gravity, as well as between stronger and weaker semantic density, to enact semantic waves in their teaching of such theoretical/abstract discipline content.

Extensive research of classroom practices showed that the use of *semantic waves* enables cumulative knowledge-building (Maton 2013; Clarence 2016; Kirk 2017), a key aspect in 'connecting the dots' of knowledge. Clarence (2016) showed that Semantics can be used by lecturers to understand how to facilitate cumulative knowledge-building using semantic waves. In the field of academic writing, Kirk (2017) demonstrated how students can be taught to use the concepts of semantic gravity and semantic gravity waves to understand what is valued and required in their writing assignments. Matruglio *et al.* (2013) used the interesting approach of temporality in classroom practice to enact semantic waves.

The extent to which students are able to enact semantic waves in discourse has been shown to play a role in achievement (Maton 2013). Research revealed that high-achieving student essays are characterized by a wider semantic range than that of low achieving essays, which often display so-called semantic flatlines – little or no movement between simpler, contextualized and more complex, decontextualized meanings (Kirk 2017). However, this depends on the questions asked or the aims of a project. Georgiou's studies (2016) in Physics education showed that students lacking experience in science (more novice learners) expressed a very limited range of semantic gravity in explanations, often remaining at the very concrete levels of stronger semantic gravity. Students with a stronger science background seem to understand that a wider semantic gravity range is needed to explain and answer certain questions. They also found that more proficient students understood which questions required a certain range for semantic gravity. However, less proficient students were found to often draw on explanations too weak in semantic gravity, thus reaching up the semantic gravity scale even when it is not necessary, revealing their lack of discernment.

Using the Semantics dimension of LCT to enact semantic waves in science education has vast potential to improve pedagogy and promote students' learning, understanding and achievement. In science lecturing, for example, lecturers may reach back to discipline content from school but also stretch toward the new complex discipline content and move between abstract theory and applications in recurrent cycles. In this type of classroom practice, knowledge is continuously transformed between relatively concrete and decontextualized meaning, as well as between simpler and more complex condensed meaning, leading to the ability to build on previous knowledge and the transfer thereof into new contexts – crucial in science education.

Scientific language is generally complex and therefore represents stronger semantic density. However, 'complexity' is a relative term and is often used simply to refer to the cognitive demand of an assessment or assignment. In contrast, 'semantic density' affords greater specificity, conceptualizing complexity in terms of the condensation of meaning within practices, where condensation refers to adding meaning to a term or practice. Maton and Doran (2017a, 2017b) distinguished between forms of semantic density and explored *epistemic–semantic density* (ESD) which deals with epistemological condensation of formal disciplinary definitions and descriptions. They offer different tools for analyzing the ESD of language at the level of individual words, word-grouping, clausing and sequencing. Epistemic–semantic density further explores the relationality of meanings. Thus, the greater the number of relations to other meanings of terms or concepts, referred to as a *constellation* of meanings, the stronger the epistemic–semantic density (Maton 2013; Maton and Doran 2017b). For example, a scientific term such as 'protein synthesis' includes actions and processes with multiple distinct parts, each with its technical meaning, and will therefore have stronger ESD.

Autonomy

The Autonomy dimension of LCT explores the degree of insulation of practices — how insulated are the parts, and how insulated are the ways that they are related together (Maton and Howard, 2018). The two organizing principles are *positional autonomy* (PA) and *relational autonomy* (RA). Autonomy is based on the assumption that any set of practices comprises both constituents (the things in the practice, i.e. concepts, ideas, artefacts, actors) and relationships among those constituents (e.g. procedures, conventions, aims).

The degree to which a constituent in a particular context is insulated from constituents in other contexts is conceptualized as *positional autonomy* – the greater the degree of insulation, the stronger the positional autonomy (Maton and Howard, 2018). In education, this is often used to distinguish between what is seen as part of, or 'inside,' a specific knowledge practice and what is not. Those things that are taken to be 'inside' a practice are defined as having stronger positional autonomy (PA+), and those considered to be 'outside' are defined as having weaker positional autonomy (PA–). For example, it can be used to analyze whether ideas are coming from within a specific topic of science (PA++), wider scientific knowledge (PA+), other academic knowledge (PA–) or everyday understandings (PA– –).

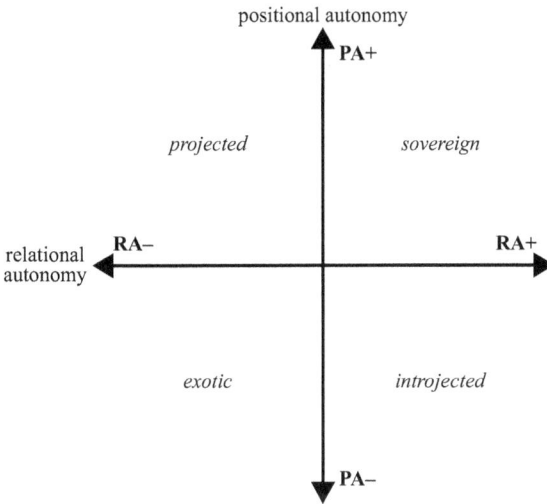

Figure 1.5 The autonomy plane (Maton and Howard 2018: 6).

The degree to which the principles governing the relations among constituents are bound by the field is conceptualized as *relational autonomy* (Maton and Howard, 2018). In education, this is generally taken as the purpose of an activity. Purposes that are taken as a legitimate part of or 'inside' a specific practice are defined as having stronger relational autonomy (RA+) than those considered 'outside.' For example, it can be used to analyze whether the ideas being taught in a science classroom are being turned to the purpose of teaching science (RA+) or towards another purpose, such as behavioural management or engagement (RA–).

Mapping positional autonomy and relational autonomy on the autonomy plane generates four principal *autonomy codes* (Figure 1.5):

- *sovereign codes* (PA+, RA+) result from strongly insulated positions and autonomous principles – PA and RA are both relatively stronger. Such practices would use, for example, 'inside' concepts to teach or research 'inside' problems, such as using the concept of equilibria in Chemistry to determine the pH of a weak acid in a Chemistry experiment.
- *introjected codes* (PA–, RA+) result when weakly insulated constituents are used for strongly bounded purposes – i.e. when things from 'outside' are used for 'inside' purposes, such as using calculus (from Mathematics) to solve problems in Physics.
- *projected codes* (PA+, RA–) result when constituents are strongly insulated, but the principles or ways in which they relate are heteronomous. Thus, what is valued arises from within the context, but it is used for other purposes, or what is 'inside' is used for 'outside' purposes; for example, when Physiology concepts are used to evaluate the validity of a health benefit claim made by the food industry.

- *exotic codes* (PA–, RA–) arise when there are weakly insulated positions and heteronomous principles. For example, knowledge from a different context is used to achieve an end that is not related to the subject in hand, such as telling a joke to get the class's attention.

Autonomy codes have the capacity to be enacted in real-world practices, such as teaching practice. It has been shown that there should be a rationale behind the materials or practices that are selected, repurposed and connected (Maton & Howard 2018, 2020). Purposeful shifts on the autonomy plane lead to so-called autonomy tours that engage and cohesively integrate different knowledge practices or content. In contrast, poor instructional design creates pathways around the plane that leave different knowledge practices or content segmented and disconnected (Maton & Howard 2018, 2020). Thus, one can use autonomy codes to design how to incorporate different knowledge practices or content, such as real-world content from other fields into science classroom pedagogy.

The layout of this volume

Given that the primary intended audience for this volume is academics who teach within a specific scientific discipline, we decided that organization according to discipline would be most helpful. Thus, we included five categories – academic support in science, physical sciences, biological sciences, mathematical sciences and science education research. If you are new to LCT, it may help to start with the section associated with your specialty first. That way you will be familiar with the knowledge content of the subject which will make the power of the LCT analysis more visible. This approach may also lower the threshold to becoming familiar with the LCT dimensions. However, once you are familiar with those chapters, we strongly recommend that you venture out of your comfort zone into different subject areas. This will both strengthen your range of understanding of the problems encountered in science education and will improve your understanding of LCT. At the end of the volume, we have included a 'how to navigate' chapter (Chapter 12) for those who are just dipping their toes into education research. We hope this chapter will help you to lower the activation energy threshold into getting going with doing your own research.

Part I of the book is potentially useful to all readers as the focus is on academic support in science (Chapter 2). The study explores the role a reflective learning portfolio in a science access course plays in enabling students to become active, self-directed and independent learners. The reflective learning portfolio interventions focus on explicitly guiding and modelling appropriate learning practices and critical reflection about learning. Karen Ellery uses LCT's Autonomy dimension to analyze the reflective learning portfolio interventions and students' responses to them.

Part II of the book focuses on the physical sciences. In Chapter 3, Christine M. Steenkamp and Ilse Rootman-le Grange focus on assessments in an introductory Physics course at Stellenbosch University in South Africa. The authors describe a detailed analysis of the Physics exam papers using *semantic density*. Their focus was on the kinds of representation i.e., graphs, diagrams, equations, etc., and the complexity of language used in these exam papers. Results revealed that some kinds of representations and some types of questions have been unconsciously omitted from their assessments.

In Chapter 4, Zhigang Yu, Karl Maton and Yaegan Doran turn their attention to different kinds of representations found in Chemistry. They carry out an in-depth study of a Chemistry textbook to reveal the levels of complexity and abstraction in operation in the diagrams. Using *epistemic–semantic density*, they develop a new method of analyzing representations in Chemistry which can be adapted to other sciences. The main aim of the chapter is to show how epistemic–semantic density can be applied to visual representations. Whilst much attention is given to symbols and nomenclature in Chemistry education, the complexity of visual representation is relatively rarely the focus of a study. Chemistry educators can tend to presume that a diagram automatically makes the content more accessible. By showing the variation in the complexity of representations in Chemistry, this chapter challenges that assumption.

In Chapter 5 Bruno Ferreira dos Santos, Ademir de Jesus Silva Júnior and Eduardo Fleury Mortim focus on high school Chemistry. They looked at the language used by the teacher, analyzing the clustering of words and phrases. Using recordings of lessons, they show the ways in which different teachers use language in the descriptions of chemical concepts. The variation is between highly dense technical language and much simpler more accessible language. The study defines various levels between these two positions using *epistemic–semantic density*. The study shows that some teachers repeatedly move between these two positions, whilst others achieve relatively little movement.

In Chapter 6, Lizel Hudson, Penelope Engel-Hills and Chris Winberg turn their attention to a Physics course presented as part of a degree in Radiation Therapy. Teaching the fundamentals of science to health sciences students who are eager to focus on patient care, is a non-trivial challenge. This chapter explores why these students may find it difficult to understand why they need to study Radiation Physics and why the subject is challenging. This chapter suggests ways in which the notion of threshold concepts can be used to make the fundamental science more accessible. This chapter also uses Specialization to make visible the challenges of teaching a subject with a very strong theoretical foundation to a cohort who are primarily interested in learning about patient care.

Part III of the book focuses on the biological sciences which here features an introductory Biology course, a senior Physiology course and a blended course comprising Anatomy and Physiology aimed at health science students.

In Chapter 7, Gabi de Bie and Sioux McKenna look at a course entitled 'Human Biology' which has developed from the amalgamation of Anatomy and Physiology courses for health sciences students. They show the ways in which integrative knowledge-building was overlooked in curriculum design resulting in a segmented course which fails to prepare students adequately for more advanced courses which draw on the foundational knowledge presented in this course. They use Specialization and Semantics in this chapter.

Chapter 8 is authored by Marnel Mouton, Ilse Rootman-le Grange and Bernhardine Uys. They explore why Biology students find it challenging to engage with complex disciplinary text from sources such as textbooks and then demonstrate their mastery of the subject matter using appropriate scientific discourse. They draw on LCT's concept of *epistemic–semantic density* (ESD) to analyze sections of the first-year and school textbooks, as well as students' written discourse from summative assessments. They show the profound variation that exists in the proficiency of the students' scientific vocabulary and language functions, as well as the discourse of the school and first-year Biology textbooks. They consequently argue for science pedagogy that would allow students time and opportunities to develop these crucial skills. Such practice may enable students to successfully engage with the subject matter and then communicate their understanding using written discourse.

In Chapter 9, M. Faadiel Essop and Hanelie Adendorff focus on using Autonomy to analyze a project-based activity in the context of an undergraduate Physiology course. The goal of the activity was to teach students how to do science as opposed to teaching them about science. Exploring what is introduced and for what purpose, using Autonomy, show the value and dangers involved in these kinds of activities. One can spend a lot of effort on marginal activities which in fact may obscure the epistemic content necessary within the subject.

Part IV of the book turns to the mathematical sciences featuring a chapter on the transition into second-year Mathematics and a chapter applicable at all levels of tertiary study focusing on mathematical knowledge.

In Chapter 10, Ingrid Rewitzky focuses on teaching Mathematics guided by the epistemic plane. It is one of the more subject-specific chapters in the book but serves as a very useful introduction to the power of the epistemic plane in making the different kinds of knowledge used in Mathematics visible in teaching. To those without some tertiary-level Mathematics, it will require a bit of digestion, but it will be well worth your time investment. This chapter is groundbreaking and will be applicable to engineering disciplines as well.

In Chapter 11, Honjiswa Conana, Deon Solomons and Delia Marshall look at the transition from first year to second year in Mathematics. At many South African institutions, there has been significant investment in improving the first-year experience, but the transition into the second year of study can prove to be a stumbling block. In this chapter, they interrogate the experiences of both students and lecturers of a particular intervention introduced to smooth this transition in Mathematics.

The final chapter in Part V is written by Margaret A.L. Blackie and is aimed at helping those new to science education research to get something of a foothold in the new terrain. The beginning of the chapter gives a brief overview of critical realism. Whilst critical realism is one theoretical framework among many, it is a useful starting point for those entering education research from a background in disciplinary research in a STEM field. This foundation is then used to situate LCT as realist sociological theory. The second part of the chapter gives some pointers on how to begin using LCT in the scholarship of teaching and learning.

Overall then, this volume provides an overview of what can be achieved using LCT in science education. Represented here are a diversity of science fields from high-level Mathematics to service courses in Biology. In addition, all the major LCT dimensions which have been developed to date and are likely to be applicable to science educators are represented here. Thus, this book provides a solid introduction to the use of LCT in science in particular and will be useful to educators and researchers across STEM fields more generally.

Note

1 For this growing body of work, see the database of LCT publications at https://legitimationcodetheory.com/publications/database/.

References

Adendorff, H. (2011) 'Strangers in a strange land–on becoming scholars of teaching,' *London Review of Education,* 9: 305–315.

Bernstein, B. (2000) *Pedagogy, symbolic control and identity: Theory, research, critique,* revised edition, Rowman and Littlefield.

Blackie, M.A.L. (2014) 'Creating semantic waves: Using Legitimation Code Theory as a tool to aid the teaching of chemistry,' *Chemistry Education Research and Practice,* 15: 462–469.

Bourdieu, P. (1988) *Homo academicus,* Stanford University Press.

Conana, H., Marshall, D. and Case, J. (2016) 'Exploring pedagogical possibilities for transformative approaches to academic literacies in undergraduate Physics,' *Critical Studies in Teaching and Learning,* 4(2): 28–44.

Clarence, S. (2016) 'Surfing the waves of learning: Enacting a semantics analysis of teaching in a first-year law course,' *Higher Education Research & Development,* 36(5): 1–14.

Georgiou, H. (2016) 'Putting physics knowledge in the hot seat: The semantics of student understandings of thermodynamics' in K. Maton, S. Hood, and S. Shay (eds) *Knowledge-building: Educational studies in Legitimation Code Theory,* Routledge.

Kelly-Laubscher, R. F. and Luckett, K. (2016) 'Differences in curriculum structure between high school and university biology: The implications for epistemological access,' *Journal of Biological Education,* 50(4): 425–441.

Kirk, S. (2017) 'Waves of reflection: Seeing knowledge(s) in academic writing,' in Kemp, J. (ed) *EAP in a rapidly changing landscape: Issues, challenges and solutions.* Proceedings of the 2015 BALEAP Conference. Reading: Garnet.

Maton, K. (2009) 'Cumulative and segmented learning: Exploring the role of curriculum structures in knowledge-building,' *British Journal of Sociology of Education*, 30: 43–57.

Maton, K. (2013) 'Making semantic waves: A key to cumulative knowledge-building,' *Linguistics and Education*, 24: 8–22.

Maton, K. (2014) *Knowledge and knowers: Towards a realist sociology of education*, Routledge.

Maton, K. and Doran, Y. (2017a) 'Condensation: A translation device for revealing complexity of knowledge practices in discourse, part 2 – clausing and sequencing,' *Onomázein*, 77–110.

Maton, K. and Doran, Y. (2017b) 'Semantic density: A translation device for revealing complexity of knowledge practices in discourse, part 1– wording,' *Onomázein*, 46–76.

Maton, K., Hood, S. and Shay, S. (2016) *Knowledge-building: Educational studies in Legitimation Code Theory*, Routledge.

Maton, K. and Howard, S. (2018) 'Taking autonomy tours: A key to integrative knowledge-building,' *LCT Centre Occasional Paper*, 1: 1–35.

Maton, K. and Howard, S. (2020) 'Autonomy tours: Building knowledge from diverse sources,' *Educational Linguistic Studies*, 2: 50–79.

Matruglio, E., Maton, K. and Martin, J.R. (2013) 'Time travel: The role of temporality in enabling semantic waves in secondary school teaching,' *Linguistics and Education*, 24 (1): 38–49.

Morrow, W. E. (2009) *Bounds of democracy: Epistemological access in higher education*. HSRC Press.

Mouton, M. (2020) 'A case for project based learning to enact semantic waves: Towards cumulative knowledge building,' *Journal of Biological Education*, 54(4): 363–380.

Mouton, M. and Archer, E. (2019) 'Legitimation code theory to facilitate transition from high school to first-year biology,' *Journal of Biological Education*, 53(1): 2–20.

Osborne, J. and Dillon, J. (2008) *Science education in Europe: Critical reflections*. A report to the Nuffield Foundation. http://efepereth.wdfiles.com/local--files/science-education/Sci_Ed_in_Europe_Report_Final.pdf.

Rootman-le Grange, I. and Blackie, M.A.L. (2020) 'Misalignments in assessments: Using Semantics to reveal weaknesses,' in C. Winberg, K. Wilmot and S. McKenna (eds) *Building knowledge in higher education: Enhancing teaching and learning with LCT*. Routledge.

Rootman-le Grange, I. and Blackie, M.A.L. (2018) 'Assessing assessment: In pursuit of meaningful learning,' *Chemistry Education Research and Practice*, 19(2): 484–490.

Sithole, A., Chiyaka, E.T., McCarthy, P., Mupinga, D.M., Bucklein, B.K. and Kibirige, J. (2017) 'Student attraction, persistence and retention in STEM programs: Successes and continuing challenges,' *Higher Education Studies*, 7(1), pp.46–59.

Steenkamp, C. M., Rootman-le Grange, I. and Muller-Nedebock, K. K. (2019) 'Analysing assessments in introductory physics using semantic gravity: Refocusing on core concepts and context-dependence,' *Teaching in Higher Education*, 1–16.

Winberg, C., McKenna, S. and Wilmot, K. (2020) *Building knowledge in higher education: Enhancing teaching and learning with Legitimation Code Theory*. Routledge.

Part I
Academic Support in Science

2 Becoming active and independent science learners

Using autonomy pathways to provide structured support

Karen Ellery

Introduction

A student's prior educational and social context can influence their success in higher education. Schooling plays a particularly important role as it not only provides a platform of academic, numeracy and knowledge discourses upon which students can build, but it also socializes them into particular academic behaviours (Mann 2008; Crozier and Reay 2011). Students from school contexts that align well with those of higher education usually find the transition into tertiary studies manageable, but those from school contexts that do not foster a critical, open-minded and curious approach towards knowledge, and self-reliant approach towards learning, will likely encounter difficulties in their tertiary studies. Conley (2008, 2010), who has done comprehensive research on university readiness by examining different socioeconomic, schooling and university contexts, provides a useful framework in which to understand preparedness of incoming students. His multidimensional model considers four interconnected aspects of university readiness: key cognitive strategies, key content, academic behaviours and contextual skills and awareness. Academic behaviours such as self-management (including metacognitive acts of monitoring, regulating and evaluating progress), and mastery of study skills and study skills behaviours (such as time and stress management, taking notes in class, prioritizing tasks), form the focus in this chapter.

In a comprehensive study by Wilson-Strydom (2015) on student capabilities required for university access and success, which uses Conley's model as a frame, academic behaviours (referred to as 'learning dispositions'; *ibid.*) are mentioned most commonly in student interviews as proving problematic in the transition to university. However, developing appropriate academic behaviours seldom forms a significant or integrated part of any higher education science curriculum, except in stand-alone, study skills-type courses.

The terms 'academic behaviours' as used by Conley (2008) and 'learning practices' as used by Ellery (2018) encompass the concepts of both study practices and learning dispositions. Study practices relate to acts such as managing time, accessing information, preparing for class, taking notes in class, consolidating notes after class, learning with understanding and can arguably be actively addressed in class. In contrast, dispositions are less easily dealt

DOI: 10.4324/9781003055549-3

with as they relate to individuals' identities. Barnett (2009: 433) defines dispositions as 'those tendencies of human beings to engage in some way with the world around them.' He further states that learning dispositions can include a will to learn and engage, a preparedness to listen, explore and hold oneself out to new experiences, and a determination to keep going forward (*ibid.*). Similarly, Wilson-Strydom (2015: 120) defines a learning disposition as both 'having the curiosity and desire for learning,' 'confidence in one's ability to learn,' 'being an active inquirer' as well as 'having the learning skills required for university study.' Whilst these definitions recognize the role of individual agency in being a successful learner, Wilson-Strydom cautions that the socio-educational context can serve to either enhance or inhibit students will and agency. I return to this idea in the discussion.

The necessity for, and challenge of, becoming and being an autonomous or independent learner in the higher education context is well recognized (see Leese 2010; Hockings *et al.* 2018; Breeze *et al.* 2020). A recent study on enabling epistemological access in a higher education science access course, drawing on Legitimation Code Theory (LCT) which helps unpack the codes or 'rules of the game' of social practice (Maton 2014), indicated that two different codes were being legitimated: a science-related *knowledge code* and a learning practices-related *knower code* (Ellery 2018). The findings indicate that if a student doesn't develop the right learning practices of being an autonomous (active, self-directed, independent, critical) learner, as required by the knower code, it becomes difficult for them to properly access, with appropriate depth and understanding, the disciplinary knowledge of the knowledge code. In response to the Ellery (2018) findings, a number of interventions under the banner of a reflective learning portfolio (RLP) were introduced into the science access course. These interventions form the object of the current study, and the question being addressed is *whether and how RLP interventions can enable development of appropriate student learning practices.*

Whilst early RLP interventions in the science access courses simply required student self-reflection, later ones obligated students to not only engage with their learning practices but also with science or disciplinary knowledge. These later interventions were based on the premise that integrative pedagogies that bring together different knowledge practices, in this case learning practices knowledge and science or disciplinary knowledge, can lead to enhanced learning. Maton (2014) and Garraway and Reddy (2017) argue that learning something in one context and being supported to use it in another can enable students to develop more connected and advanced forms of meaning than when learning occurs in a single context only. Empirical work by Virtanen and Tynjälä (2019) on learning of generic skills, by Garraway and Reddy (2017) on learning about writing and presenting and Maton and Howard (2018) on learning disciplinary knowledge support this contention. The concepts of *autonomy codes* from LCT, which show how different knowledge practices can be successfully brought together (Maton 2014; Maton and Howard 2018, 2021a, 2021b), were therefore used as the analytical tool in this study to ascertain the role of the RLP in enabling effective learning.

Conceptual and analytical framework

LCT is a conceptual and analytical framework used to research social practices by making visible the 'codes' or 'organizing principles' that underpin a particular practice (Maton 2014). This takes us beyond simple descriptions to an explanatory account of any practice. The codes of social practices are governed by the primary participants and may be quite overt, but more often than not, they are unarticulated and form an implicit part of the norms and values of that practice. LCT provides the 'tools' to help expose the hidden codes of a practice through the use of a number of dimensions. Of interest here is the Autonomy dimension of LCT as it allows us to see how different knowledge practices can be, and are, integrated in classroom practice.

The Autonomy dimension is based on the premise that any social practice comprises *constituents* that are *related* in particular ways (Maton and Howard 2018). The constituents may be people, ideas, objects, institutions, etc., and these constituents occupy a certain position in the context. This gives rise to the first organizing principle of autonomy analysis: *positional autonomy* (PA). If the constituents are strongly positioned or tightly insulated in the context, this represents stronger positional autonomy (PA+), but if the constituents are relatively weakly delimited and drawn from constituents in other contexts, this represents weaker positional autonomy (PA−).

The second organizing principle in autonomy analysis is that of *relational autonomy* (RA). It refers to how constituents in a context are *related* to one another through specific procedures, implicit conventions, stated aims, formal rules, etc. (Maton and Howard 2018). If procedures and rules are specific and autonomous to that practice, this represents relatively stronger relational autonomy (RA+). However, if rules and procedures are drawn from or shared with other practices, they are heteronomous, which represents weaker relational autonomy (RA−).

As in all LCT dimensions, the two organizing principles can be plotted on a two-dimensional plane. The autonomy plane gives rise to four principal autonomy codes (Figure 2.1). *Sovereign codes* (PA+, RA+) refer to practices with strongly insulated constituents and autonomous principles. Sovereign codes are key in this dimension as it reflects what constituents and what purpose/s are constitutive, or the *target*, of that particular practice (Maton and Howard 2018: 10). In other words, sovereign codes are the starting point for expressing target (or internal) constituents and target (or intrinsic) purposes that represent the practice, and any other constituents or purposes are considered non-target (or external). *Exotic codes* (PA−, RA−) therefore represent practices drawing on non-target constituents for non-target purposes. *Introjected codes* (PA−, RA+) refer to practices using non-target constituents for target purposes. *Projected codes* (PA+, RA−) represent practices using target constituents for non-target purposes.

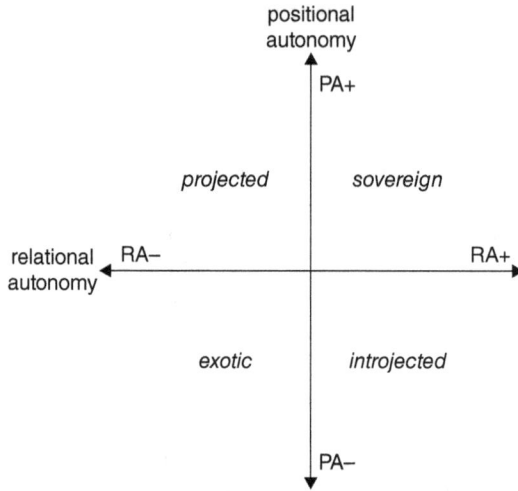

Figure 2.1 The autonomy plane (Maton and Howard 2018: 6).

For example, if the target constituents in a Biology lesson are the concepts of energy flows and trophic levels, this represents the measure of stronger positional autonomy (PA+), and if the target purpose is learning such concepts, this represents the measure for stronger relational autonomy (RA+). Classroom practice that focuses on these concepts for this purpose would therefore be located within a sovereign code. However, if non-target mathematical calculations to quantify energy flows are used, representing weaker positional autonomy (PA−), but still for the target purpose of understanding energy flows, representing stronger relational autonomy, (RA+), this would be located in an introjected code. Likewise, if non-target mathematical calculations representing weaker positional autonomy (PA−) are used for the non-target purpose of understanding Mathematics rather than the Biology concepts, representing weaker relational autonomy (RA−), this would be located in an exotic code for this particular Biology lesson. In a Mathematics classroom, however, mathematical knowledge would be the target content, and learning such knowledge would be the target purpose.

An empirical analysis of any social practice over time allows us to either *position* the practice on the two-dimensional autonomy plane and identify *pathways* between autonomy codes (Maton and Howard, 2018). Through plotting positions and/or pathways for different classroom practices in a school context, Maton and Howard (2018) suggest that *stays* in one code and *one-way trips* from one code to another will likely result in different knowledge practices remaining isolated and segmented. However, classroom practices that move back and forth between and across to other codes, resulting in different knowledge practices being selected, repurposed and reconnected (*ibid.*: 33), can play an important role in integration and therefore knowledge-building. The current study draws on these concepts to examine pathways taken in the RLP interventions and whether they contribute to enabling effective knowledge-building and learning in the sciences.

Context of study

The study described here takes place in the aforementioned science access course that forms part of an additional, supported year in an extended degree programme (EDP) at a small South African University. Students can proceed from the EDP into any science degree at the university. EDP students are generally first-generation university learners who attended under-resourced and poor-quality schools. The purpose of the year-long, multidisciplinary, science access course, which includes Physics, Chemistry, Life Sciences (Human Kinetics and Ergonomics) and Earth Sciences disciplinary content, is to not only enable access to scientific concepts and literacies but to also enhance the development of learning competencies and practices appropriate in higher education science (SESP Review Report 2011). Learning competencies and practices have relatively recently been addressed formally in the science access course through the introduction of a RLP.

The purpose of the RLP was to formalize student reflection on their own learning, to model certain learning activities in the classroom and to provide more opportunities for dialogue with students about their own learning. This was in recognition that students needed to become much more independent in their approach to their studies as well as learn to develop good conceptual understanding, which is encapsulated as their becoming and being autonomous learners (Ellery 2018). This was articulated as such in the RLP introductory document:

> To be the right learner at university you need to participate actively both in and out of class, be willing to be challenged, develop good conceptual understanding and engage at a high level with material (i.e. comprehension rather than memorization), seek help when needed, and work independently (without being told when to work). In other words – you need to become and be an autonomous learner.

At each RLP intervention, student activities were guided by a task handout. Students usually engaged interactively with handout questions, the facilitator and each other but provided individual written responses. It was emphasized that in reflections there were no 'right' answers, and to encourage free writing, their reflective responses in particular did not have to be 'academic' or grammatically correct. Their written responses were commented on individually by the facilitator in a dialogic and engaging manner. A low-stakes summative mark was awarded for their reflections, but only upon final submission of the entire portfolio.

Methodology

The concepts of autonomy analysis are abstract and far removed from the empirical data that make up any practice, requiring a means of bridging this 'discursive gap' (Bernstein 2000: 32) to show how the concepts are realized within a specific object of study. To achieve this, the *translation device* devised by Maton and Howard (2018) was used to analyze pedagogic practices associated with RLP interventions in the science access course. This device draws on the already mentioned target content (PA+) and target purpose (RA+) and non-target content (PA−) and non-target purpose (RA−). When analyzing each intervention, the specific target content and purpose were identified, and any content or purpose that strayed from this was considered non-target.

Three sources of data were used: (a) tutorial handouts for all 14 interventions, which consisted of stated aims and objectives and various tasks to guide students' activities and reflections, (b) written responses of all 44 students in the class to the tasks and (c) semi-structured interviews with 17 volunteers. The interviews focused specifically on probing specific students' written responses to tasks for further clarification, as well as ascertaining more broadly students' level of engagement with, and perceived value of, the RLP.

For each pedagogic intervention, the stated aim in the tutorial handout indicated target content and purpose. Each subsequent task was then categorized as either target or non-target content and purpose. These data were used to code the 'expected' position/s or pathway/s on the autonomy plane for a particular intervention. Student responses were similarly coded, to indicate 'observed' position/s or pathway/s on the autonomy plane, which did not always match what was intended by task questions. Interview data provided insights beyond the written responses and helped explain observed responses.

I am involved in the science access course in question through being both an observer of disciplinary lectures and a facilitator in some of the RLP interventions. As indicated by Chavez (2008), being an insider researcher usefully offered knowledge of the context and legitimacy and rapport with participants, but it also required me to be critically reflexive, member check all interviews and consult with colleagues on my data interpretation.

Results: Autonomy pathways in RLP interventions

The 14 RLP interventions could be categorized broadly into one of three task-types. The first task-type involved mainly *self-reflection* relating to students' expectations of university and personal goals and attributes. The second task-type involved guided *modelling of appropriate learning practices* and included topics such as time management, taking lecture notes in class, preparing for a practical, consolidation of a lecture, learning for multiple-choice and essay tests, developing test questions and active learning. The third task-type was *guided reflection on performance in assessment*, which required students to reflect on their test learning and use of essay criteria.

Whilst each tutorial intervention produced slightly different expected and observed autonomy positions and pathways, a 'typical' example of each task-type is presented here.

Self-reflection: Strengths and weaknesses as a learner

In the first week of term, students were asked to describe their academic goals at university and to consider personal attributes that would either enable or constrain them in achieving their goals. They were also asked to make suggestions on how to overcome constraining attributes. The target content of the tutorial was therefore personal attributes that enable or constrain students in reaching their goals (PA+), and the target purpose was for students to acknowledge their strengths and find solutions to attributes that could be a hindrance in achieving their goals (RA+), which is located in a sovereign code. Since the intervention required students to draw on self-reflections rather than disciplinary knowledge or practices, the intention of this intervention was for them to be positioned and work in a sovereign code (solid line a, Figure 2.2). What was 'observed,' however, based on analysis of students' tutorial answers, did not always match the lecturer's intention (dashed lines, Figure 2.2).

Based on their written reflections, students seemed to easily identify enabling and constraining attributes about themselves. Enabling attributes related to being a hard worker, not giving up easily, enjoying participating in academic activities, asking questions, loving science, as well as being curious, assertive, motivated, goal-driven, outspoken and determined. Constraining attributes mentioned related to getting bored easily, being easily distracted, being too shy to ask questions in class, doubting themselves and their

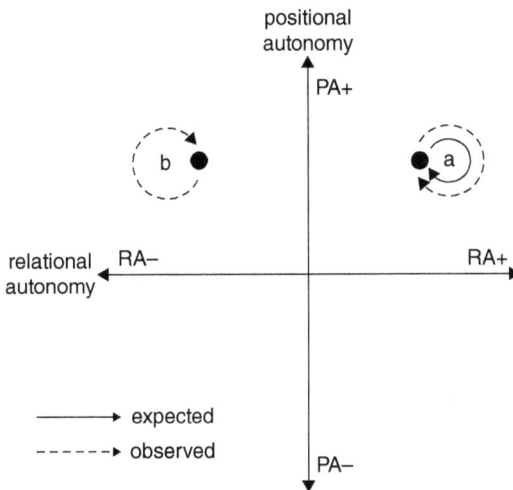

Figure 2.2 Autonomy pathways for self-reflection intervention.

capabilities, being careless in their work and not managing their time. Various solutions were presented: finding a group to work with if they are too shy to ask in class, working in the library to avoid distractions, using study timetables and working daily to overcome poor time management and procrastination. These students were both working with the target content of identifying personal attributes as enabling or constraining (PA+) for the purpose of considering ways in which they could overcome any that may prevent them reaching their goals at university (RA+). This means they were positioned in a sovereign code for this tutorial (dashed line a, Figure 2.2), as was intended.

However, some students simply listed what they considered positive and negative attributes but provided no solutions. They were therefore still working with the target content of attributes (PA+) but were not achieving the target purpose of finding solutions (RA–). These students were there-fore positioned and working in a projected code (dashed line b, Figure 2.2). It was striking in the interviews that many students valued this self-reflective approach, primarily because they found it motivating and affirming to think of their own strengths. As one mentioned: 'It was a motivation for us ... I wouldn't have thought to think of my strengths [on my own]' (KJ).

The other striking feature was the relief students felt in articulating possi-ble constraints to learning, partly because they felt the lecturer would know them better. As one said: 'Those reflections are for my lecturer to read ... they can understand the kind of student I am' (MK). For some, this was motivating, as they felt some compulsion to follow through on their own reflections. In this regard, one stated: 'I know you read what I said I would do, so now I know I must do it, you know, follow through' (AD). For oth-ers, the relief was contingent on lecturer feedback and suggestions which, to their frustration, wasn't always satisfactory. One interviewee stated: 'Maybe if you guys would have reached out again — said this is the problem you encounter ... how can we help you?' (LP).

Three of the 17 interviewees indicated they did not find the self-reflection tasks useful as they were at a loss as to how to go about making the necessary changes. As one said: 'I was just like speaking them but didn't know how to go about changing' (MV). Another stated: 'It did help me know where I stand but not exactly move where I want to be' (EQ). These students were working in a projected code and clearly needed closer guidance, in terms of finding solutions, to move them into a sovereign code.

Modelling learning practices: Consolidating a lecture

In the fifth week of the academic year, a Physics lecturer presented a 90-minute double lecture in the science access course on 'measurement in science' that focused on aspects of precision, uncertainty and the use of sig-nificant figures. The lecturer went through the material thoroughly, inter-acted constantly with the students and wrote main ideas and all relevant information on the board. Nonetheless, the concept of significant figures was new for the majority of students and conceptually challenging. Students were expected to write their own notes for this lecture.

The process of independently consolidating work to ensure better learning and understanding is encouraged in the science access course. Initially, the process is modelled in class tutorials, but students are later expected (and reminded) to do this without input from staff. Modelling consolidation of the 'measurement in science' lecture was done through asking students to identify the main topic and sub-topics of the lecture, to develop a 'big picture' understanding of the lecture. Students had with them their lecture notes from class to achieve this goal. The final hand-in product was a summary of the lecture in any form they wished, with most choosing to do a mind-map.

The target content of the tutorial was the concept of reviewing a lecture through identifying the main topic and sub-topics, signifying stronger positional autonomy (PA+). The target purpose was for students to learn how to identify the main topic and sub-topics to develop a 'big picture' of a lecture, signifying stronger relational autonomy (RA+). The concept of consolidating a lecture through the identification of key topics is a generic one that is located in a sovereign code (PA+, RA+). Nonetheless, in order to achieve this, it is necessary to draw on non-target science content, representing weaker positional autonomy (PA–), but still for the target purpose of learning to identify relevant topics, representing stronger relational autonomy (RA+). This tutorial, therefore, expected students to be positioned and work in an introjected code by drawing on the non-target lecture content of measurement in science for lecture consolidation (solid line a, Figure 2.3).

A few students set about the task competently, and their actions matched the expectations of the tutorial. Using their comprehensive lecture notes,

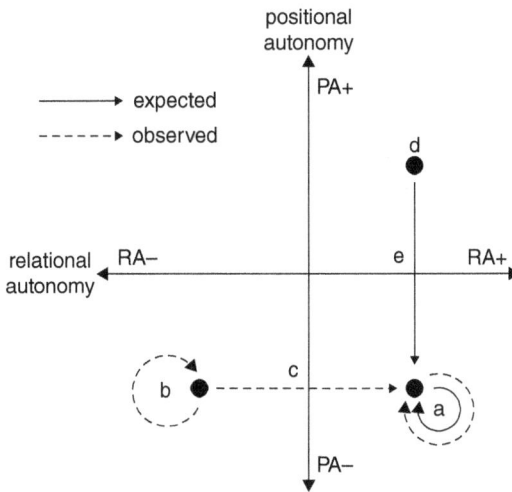

Figure 2.3 Autonomy pathways for modelling consolidation of a lecture.

which represent non-target content (PA–), they identified the main topic as well as the relevant sub-topics, which represent target purpose (RA+), with relative ease. They, therefore, occupied a position in an introjected code completing this task (dashed line a, Figure 2.3).

In contrast to the actions of these few students, the majority had inadequate lecture notes for the purpose of consolidating a lecture in this way. Since it was early in the year, students were still struggling to listen actively, distinguish relevance and record appropriate information in a lecture context. Consequently, the facilitator intervened and took 45 of the 90 minutes to answer student questions on lecture content. This additional in-class intervention, therefore, focused on non-target science content, signifying weaker positional autonomy (PA–), for the purpose of understanding non-target science content, signifying weaker relational autonomy (RA–), and represents working in an exotic code. Some students spent the rest of the tutorial simply engaging in non-target science content (weaker positional autonomy; PA–) for the non-target purpose of understanding content (weaker relational autonomy; RA–). This resulted in their being positioned in an exotic code for the entire tutorial, and therefore not meeting its aims (dashed line b, Figure 2.3).

However, after the in-class intervention, a number of students worked with non-target science lecture content (weaker positional autonomy, PA–) to complete the target purpose of identifying relevant sub-topics and details (stronger relational autonomy, RA+). This represented a trip from an exotic code to an introjected code (dashed line c, Figure 2.3).

The tutorial tasks required students to be positioned in an introjected code. However, the target content and purpose of the tutorial are that students recognize the generic nature of lecture consolidation through identification of key topics (PA+), and that they effectively use this approach when working independently (RA+). Such recognition would position them in a sovereign code (dot d, Figure 2.3), although they would need to move into an introjected code to complete the consolidation task (solid line e, Figure 2.3). However, interviews indicated that, because their understanding of lecture material had been so poor, students had viewed this particular RLP intervention as an 'extra class' for learning lecture content. As a result, few consolidated lectures in this way as a generalized practice, highlighting the need to better articulate and emphasize target content and purpose of the interventions. Some interviewees indicated that lecture consolidation was difficult to do independently and that more active modelling in class and closer support were needed. As one student put it: 'I face problems when consolidating a lecture without having questions [from the lecturer]' (LP). A more carefully structured approach was used effectively in the following task-type.

Guided student reflection: Learning process based on performance in a test

The third term test for the Human Kinetics and Ergonomics component of the science access course was about human movement. During a RLP intervention, students were required to reflect on their test learning using three sets of documents: their marked test, their lecture notes (PowerPoint slides annotated by themselves during lectures) and the tutorial handout guiding their reflections. As indicated by the aims and objectives of the tutorial, the target content for this tutorial was students' learning practices both in and out of class (PA+), and the target purpose was to consider appropriateness of, and ways of improving, learning practices (RA+).

The tasks in this tutorial guided them carefully through a set of reflections. In order to make students engage both with lecture notes and test questions simultaneously, the first task required students to examine a particular PowerPoint slide from a lecture. This had a graph showing the length-tension (force) relationship in muscles that was similar to the test graph. Students were asked to describe how they had engaged in learning about and understanding the graph, which represents target content and therefore signifies stronger positional autonomy (PA+). They were also required to consider the appropriateness of their learning approach, which represents the target purpose of guiding students to reflect on effectiveness of learning, signifying stronger relational autonomy (RA+). This first activity represents positioning in a sovereign code (solid line a, Figure 2.4). The majority of students did indeed work in a sovereign code as indicated by their tutorial comments that they needed to, among others, (a) not ignore graphs in their learning, (b) engage more with complex work and (c) develop proper understanding rather than memorizing (dashed line a, Figure 2.4).

The next task related to graphing principles. Since the test question required a description of a graph, students were asked what 'rules' needed to be applied to such a description. The content of the intervention was no longer about learning practices but instead about non-target 'rules' for graph

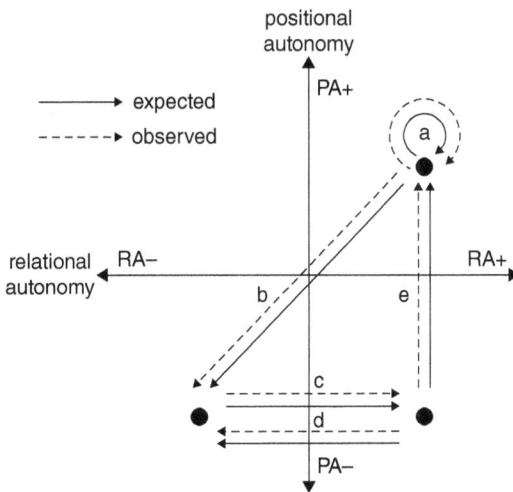

Figure 2.4 Autonomy pathways for guided reflection on a test.

description such as naming the variables, indicating how the variables relate to one another, describing the shape of the graph and referring to actual values of variables where possible. This represents weaker positional autonomy (PA–). Similarly, the purpose was no longer to reflect on learning practices, but rather to ensure non-target learning of graph description 'rules,' signifying weaker relational autonomy (RA–). By engaging with this task, students were therefore taking the expected trip from a sovereign code to an exotic code (solid and dashed lines b, Figure 2.4).

What followed was a series of tasks that involved pathways back and forth between exotic and introjected codes. Each time students were asked to consider non-target science from the test questions (such as 'rules' for describing a graph or information on gross muscle structure) which signifies weaker positional autonomy (PA–), for the target purpose of considering the effectiveness of their learning of such science (what they learned, how well, how could they improve), signifying stronger relational autonomy (RA+), they would travel from an exotic code to an introjected code (solid and dashed lines c, Figure 2.4). They would then be asked new non-target science associated with the next test question, representing weaker positional autonomy (PA–), for the purpose of learning the non-target science, representing weaker relational autonomy (RA–). This would require a trip from an introjected code back to an exotic code (solid and dashed lines d, Figure 2.4). This iterative approach resulted in many trips between these two codes. Areas for learning improvement identified in student reflections related to their (a) better relating lecture content with the question being asked, (b) understanding and using terminology correctly, (c) being able to apply ideas in different contexts and (d) not working superficially.

The final tutorial task required students to provide general comments on their engagement in and out of class, for achieving success in the Human Kinetics and Ergonomics section of the course. Students were therefore being asked to consider their learning engagement and practices (representing target content; PA+) and how effective they were (representing target purpose; RA+). This represents a final trip from an introjected code back into a sovereign code (solid and dashed lines e, Figure 2.4). Most students provided good critical reflections. One, who achieved 78% for the test, indicated learning practices that were clearly working for her:

> During the lecture I take down notes of the things I consider important especially those that are not on the handout. Then on Tuesdays I then review the notes together with the handouts, and lastly I revise what I did.
>
> (VN)

Another, who had failed the test, identified that she needed to work more independently and thoroughly: 'Need to look at my work at least each night if I get the chance. Try to learn to know and understand so that I can be able to explain in my own words' (MS).

Students found this carefully guided approach, as well as the integration of learning practices knowledge with science knowledge, effective for critical reflection on learning practices. As one said:

> Usually [when I get a test back] I just think, uh, maybe it will improve next time. But when we had to, to actually look at my answer and think about it, I could not escape [laughs] ... when my graph answer was not good, at that moment I realized I was just learning superficially, you know, with not understanding.
>
> (SQ)

This approach was acknowledged to be useful for learning disciplinary content as well. One student stated: 'I actually learned a lot about HKE doing this [RLP tutorial intervention]' (MK). Students also admitted that reflecting on marked material made them engage well with the task. As one interviewee commented: 'Marks are the most useful form of pressure to change' (LZ).

Discussion

Reflection, a reasoning process that helps make meaning of experiences, is increasingly recognized as key for developing lifelong learners (Boud *et al.* 2013). This study considers whether and how interventions that guide students' reflections can enable development of appropriate learning practices in a higher education context. Whilst the data in this study does not allow me to claim that student learning practices did indeed improve, it does reveal two key findings: practices that either: (a) integrate diverse knowledge practices in a carefully structured manner or (b) create a positive learning environment are likely to be effective in enabling such learning. These are discussed in turn below.

A structured approach to integrating diverse knowledge practices

The evidence suggests that integrating learning practices knowledge with disciplinary knowledge, through carefully structured tasks that guide students to take repeated trips between different autonomy codes, can be effective in ensuring students reflect effectively on their own learning practices and that this can lead them to becoming more active and independent learners. This appears to be most effective when reflections are in response to marks obtained in disciplinary-related assessment tasks, which supports the notion of Gibbs (1999) and Biggs (2002) that levels of student engagement are mostly driven by mark-associated tasks.

When a student practice does not match the intended practice, this is referred to in LCT terms as a *code clash* (Lamont and Maton 2008). Mapping of autonomy codes is particularly useful in highlighting such clashes and indicates where future attention in interactions can be focused. In this study, the code clash in the 'self-reflective' tasks indicated a need for better dialogue with students in order to encourage and guide their seeking ways to overcome hindrances to achieving their goals. The code clash in the 'modelling learning practices' tasks occurred when students stayed focused on learning disciplinary content in an exotic code when they were instead meant to return to thinking about their learning practices in an introjected code. As one interviewee indicated, when there are competing demands between disciplinary knowledge and learning practices knowledge, the former is favoured: 'I guess I focused on … [disciplinary knowledge] … because that is what I thought I needed to know, you know, for marks' (XG). This highlights the need for ensuring good disciplinary knowledge before attempting RLP interventions that draw on such knowledge. Furthermore, as suggested by Maton and Howard (2018), a deliberate move away from working with specific disciplinary knowledge (in an exotic code or introjected code) to more general learning practices (in a sovereign code) at the end of an RLP intervention, may ensure better uptake of appropriate learning practices – to ensure better learner autonomy. This likely would have been useful in the 'modelling learning practices' intervention and appeared effective in the 'guided student reflection' intervention.

There is an obvious tension in advocating for close guidance when the purpose of the task is to enhance learner autonomy. Nonetheless, this study clearly indicates a need for structured guidance, particularly with students in their first year of tertiary study, without which reflections will likely be poor and development of desired learning practices such as consolidating a lecture or using criteria effectively may be beyond the reach of some. As Hockings *et al.* (2018: 156) suggest, a 'dependency weaning' is needed, where there is initial guidance but a scaffolded reduction in support.

Creation of a positive learning environment

The importance of the early self-reflective interventions, where it was expected that students would be positioned and stay in a sovereign code, was not in the integration of diverse knowledge practices, but rather in terms of motivating and affirming students. This incorporates the affective dimension and takes into account the role of personal attributes and dispositions and their influence on student learning.

There is growing evidence that students' learning dispositions significantly influence how they engage with new learning opportunities (Shum and Crick 2012: 2). However, it can be argued that learning dispositions are not necessarily fixed or certain but are instead dependent on the context in which they are enacted (Sadler 2002; Wilson-Strydom 2015). This sociocultural view recognizes that whilst exercising agency to become a different kind of learner is individually negotiated and results in personal re-orientation, the educational context can serve to foster or inhibit such agency (Wilson-Strydom 2015: 140). The value of the RLP in this study is that it appears to serve to create a positive learning environment that both acknowledges students' strengths and offers a safe space in which a positive dialogic relationship can be developed between staff and students. These are discussed in turn.

Whittaker (2008) and Schreiner and Hulme (2009) suggest that a successful transition in a new context should be rooted in building on the values and strengths (and skills and knowledge) of what students bring with them to the context. This represents a move away from the deficit student model (Smit 2012; Schreiner and Hulme 2009) to one in which pedagogy is responding to student needs by providing conditions that enhance transition. The strengths-based approach as advocated by Schreiner and Hulme (2009: 74–5) indicates there is likely increased student engagement in the learning processes, a positive sense of academic self-efficacy and higher levels of perceived academic control which includes belief that they have the attributes necessary for academic success. The interviews in this study point to students feeling positive and more confident about their learning practices after engaging in the 'self-reflection' intervention.

Christie *et al.* (2008) and Tett *et al.* (2017) comment on how positive staff-student relationships make a significant difference to how students cope, largely through building confidence. Of particular note in the interviews in this study is the value students place in their positive relationship with staff, both through the RLP being a 'safe space' to articulate their concerns and insecurities about learning at university, as well as through conversational and encouraging feedback comments. Feedback can be used to support learning and Paget (2001) comments specifically on the key role a trusted respondent can play in enabling deep and critical reflections and even changing practices. Dialogic feedback was used by staff in the RLP to both encourage and suggest, as the following example from the RLP intervention on time management shows:

> Good description of time management VN! Making summaries is also useful. Have you considered setting questions for yourself to answer? I hope you will also start asking yourself other more challenging questions that will lead you to think more deeply about the subject matter at hand. What I have in mind here is questions that start with 'What if ....'. You never know where that will take you!

One interviewee captured, in general, their response to and engagement with the feedback:

> It [feedback] makes me feel good ... because I really doubt myself sometimes. And then I am encouraged to try even harder ... I look at the problem and focus on the problem where the lecturer says I may be stumbling.
>
> (SN)

Concluding comments

Whilst the focus in this study has been on science students in a higher education access course, I contend the findings have broader implications for the sector. Globalization and university massification processes have resulted in a student body that hails from a range of social, cultural and educational backgrounds, some of which have not prepared students well for tertiary study. This has led to increasing calls for socially just pedagogies that meet the needs of all students within the system (see Arum *et al.* 2012; Wilson-Strydom 2015). This chapter responds to this call by showing that RLP interventions that make explicit expectations and also guide and model appropriate learning practices and critical reflections have the potential to play a role in enabling students becoming and being autonomous learners in a high education context. Of particular interest in this chapter is the use of LCT autonomy codes as an analytic tool, as it underscores the effectiveness of integrating learning practices knowledge with disciplinary knowledge to ensure good student engagement and reflection in terms of their own role in becoming effective learners. It also indicates, through the concept of code clashes, where tutorial interventions could be improved. The importance of positive and appropriate socio-cultural-education conditions, in assisting students enacting their own learner agency, is also highlighted in this study.

Acknowledgements

The following are thanked for their contribution: Dr Nkosinathi Madondo for fruitful discussions on the RLP; anonymous reviewer for very useful autonomy codes insights.

Disclosure statement

There is no potential conflict of interest in this study. Ethical permission was obtained from Rhodes University Ethics Committee (SCI2018/001)

Funding

The study was supported by funding from Rhodes University Research Committee.

References

Arum, R., Gamoran, A. and Shavit, Y. (2012) 'Expanded opportunities for all in global higher education systems,' in L. Weis and N. Dolby (eds) *Social Class and Education: Global Perspectives*, Routledge.

Barnett, R. (2009) 'Knowing and becoming in the higher education curriculum,' *Studies in Higher Education*, 34(4): 429–440.

Bernstein, B. (2000) *Pedagogy, Symbolic Control, and Identity: Theory, Research, Critique*, Rowman and Littlefield.

Biggs, J. (2002) 'Aligning the curriculum to promote good learning,' paper presented at Constructive Alignment in Action: Imaginative Curriculum Symposium,' LTSN Generic Centre, November. Available at: http://www.qub.ac.uk. (Accessed 3 February 2018).

Boud, D., Keogh, R. and Walker, D. (2013) *Reflection: Turning Experience into Learning*, Routledge Falmer.

Breeze, M., Johnson, K. and Uytman, C. (2020) 'What (and who) works in widening participation? Supporting direct entrant student transitions to higher education,' *Teaching in Higher Education*, 25(1): 18–35.

Chavez, C. (2008) 'Conceptualizing from the inside: Advantages, complications, and demands on insider positionality,' *The Qualitative Report*, 13(3): 474–494.

Christie, H., Tett, L., Cree, V.E., Hounsell, J. and McCune, V. (2008) 'A real rollercoaster of confidence and emotions: Learning to be a university student,' *Studies in Higher Education*, 33(5): 567–581.

Conley, D.T. (2008) 'Rethinking college readiness,' *New Directions for Higher Education*, 144: 3–13.

Conley, D.T. (2010) *College and Career Ready. Helping all Students Succeed beyond High School*, Jossey-Bass.

Crozier, G. and Reay, D. (2011) 'Capital accumulation: Working-class students learning how to learn in HE,' *Teaching in Higher Education*, 16(2): 145–155.

Ellery, K. (2018) 'Legitimation of knowers for access in science,' *Journal of Education*, 71: 24–38.

Garraway, J. and Reddy, L. (2017) 'Analyzing work-integrated learning assessment practices through the lens of autonomy principles,' *Alternation*, 23(1): 285–308.

Gibbs, G. (1999) 'Using assessment strategically to change the way students learn,' in S. Brown and A. Glasner (eds) *Assessment Matters in Higher Education*, SRHE and Open University Press.

Hockings, C., Thomas, L., Ottaway, J. and Jones, R. (2018) 'Independent learning – What we do when you're not there,' *Teaching in Higher Education*, 23(2): 145–161.

Lamont, A. and Maton. K. (2008) 'Choosing music: Exploratory studies into the low uptake of music GCSE,' *British Journal of Music Education*, 25(3): 267–282.

Leese, M. (2010) 'Bridging the gap: Supporting student transitions into higher education,' *Journal of Further and Higher Education*, 34(2): 239–251.

Mann, S.J. (2008) *Study, Power and the University*, Open University Press and McGraw Hill.

Maton, K. (2014) *Knowledge and Knowers: Towards a Realist Sociology of Education*, Routledge.

Maton, K. and Howard, S.K. (2018) 'Taking autonomy tours: A key to integrative knowledge-building,' *LCT Centre Occasional Paper 1*, 1–35.

Maton, K. and Howard, S. K. (2021a) 'Targeting science: Successfully integrating mathematics into science teaching,' in Maton, K., Martin, J. R. and Doran, Y. J. (eds) *Teaching Science: Knowledge, Language, Pedagogy*, Routledge, pp. 23–48.

Maton, K. and Howard, S. K. (2021b) 'Animating science: Activating the affordances of multimedia in teaching,' in Maton, K., Martin, J. R. and Doran, Y. J. (eds) *Teaching Science: Knowledge, Language, Pedagogy*, Routledge, pp. 76–102.

Paget, T. (2001) 'Reflective practice and clinical outcomes: Practitioners' views on how reflective practice has influenced their clinical practice,' *Journal of Clinical Nursing*, 10: 204–214.

Sadler, R.D. (2002) 'Learning dispositions: Can we really assess them?' *Assessment in Education: Principles, Policy and Practice*, 9(1): 45–51.

Schreiner, L. and Hulme, E. (2009) 'Assessment of students' strengths: The first step to student success,' in B. Leibowitz, A. van der Merwe, and S. van Schalkwyk (eds) *Focus on First-year Success: Perspectives Emerging from South Africa and Beyond*, Sun Press.

SESP (Science Extended Studies Programme) Review Report (2011) *Curriculum Review: Science Extended Studies Programme, Rhodes University*. Unpublished Report, Rhodes University, Grahamstown.

Shum, S.B. and Crick, R.D. (2012) 'Learning dispositions and transferable competencies: Pedagogy, modelling and learning analytics,' paper presented at *2nd International Conference on Learning Analytics and Knowledge*, Vancouver, 29 April–2 May.

Smit, R. (2012) 'Towards a clearer understanding of student disadvantage in higher education: Problematizing deficit thinking,' *Higher Education Research and Development*, 31(3): 369–380.

Tett, L., Cree, V.E. and Christie, H. (2017) 'From further to higher education: Transition as an on-going process,' *Higher Education,* 73(3): 389–406.

Virtanen, A. and Tynjälä, P. (2019) 'Factors explaining the learning of generic skills: A study of university students' experiences,' *Teaching in Higher Education*, 24(7): 880–894.

Whittaker, R. (2008) *Quality Enhancement Themes: The First Year Experience. Transition to and during the First Year*, The Quality Assurance Agency for Higher Education.

Wilson-Strydom, M. (2015) *University Access and Success: Capabilities, Diversity and Social Justice*, Routledge.

Part II
Physical Sciences

3 Improving assessments in introductory Physics courses

Diving into Semantics

Christine M. Steenkamp and Ilse Rootman-le Grange

Introduction

The discipline of Physics constitutes a highly 'hierarchical knowledge structure' (Bernstein & Solomon 1999) in which interrelated core concepts build on one another. Each core concept is applicable to a vast range of real-world contexts, though understanding and modelling of most real-world problems require the combination of several concepts. The ability to transfer knowledge to different contexts is critically important in Physics education where students are required to not only master the hierarchical knowledge structure of Physics but also apply the core concepts to different parts of the curriculum, to real-life scenarios and in future unfamiliar work environments (Laverty *et al.* 2016).

In introductory Physics modules, the priority is not only the success of the students in the module but also their degree of preparation for subsequent study. The challenge is to cultivate learning that facilitates the ability to 'transfer knowledge across contexts and build knowledge over time' (Maton 2009: 45). The concept of learning that facilitates transfer has been incorporated into the term 'cumulative learning,' highlighting that transfer is essential for students to build new knowledge on previously acquired knowledge (Maton 2009: 43). Cumulative learning also enables students to apply their knowledge in new contexts (Kilpert and Shay 2013).

The concept of 'transfer' originated in the field of Psychology (Nokes-Malach and Mestre 2013; Barnett and Ceci 2002). Transfer has been studied extensively in science disciplines (Lobato 2006) including Physics (for example, Finkelstein 2005). In a review of Physics education research, Docktor and Mestre (2014: 31) highlighted research on cognition as one of the prominent research directions, with 'learning and transfer' as a focus. In first-year Physics, cumulative learning means that students must know the core concepts in the curriculum and develop the ability to apply these in different contexts. Observation shows that this approach to learning is unfamiliar and difficult to first-year students (Walsh *et al.* 2007).

DOI: 10.4324/9781003055549-5

In an ongoing study, we are investigating the application of the Semantics dimension of Legitimation Code Theory (LCT) (Maton 2014b) as a framework for analyzing our educational practice in introductory Physics. The Semantics dimension was chosen as it offers a distinction between context-dependence and complexity, which are expected to be important factors in assessment questions. The study focuses on the analysis of questions in summative assessments of the first introductory Physics module in a three-year Bachelor of Science degree. The first reason for the focus on assessment questions is that 'assessment always acts as an intervention into student learning' (Bearman *et al.* 2016: 547) since assessment communicates in the most concrete way to students what they are expected to learn (Brown and Knight 1994; Laverty *et al.* 2016). The second reason is that it is very challenging to develop cumulative learning in a real academic environment where teaching styles vary, staff may change and it is difficult to enforce uniform teaching practice. However, the setting of summative assessments (tests and exams) is a regulated process involving all lecturers responsible for the module and the internal moderator, and it is possible to reach an agreement on how this process is to be executed and to implement changes. As assessment is a high-stakes activity for both students and teachers, we expect that a better understanding of what is assessed will influence both teaching and learning.

We consider this study important for the improvement of our own teaching and assessment practices. We as Physics teachers are comfortable with the language of Physics but often blind to the hurdles that exist for the students in acquiring not only the explicit knowledge but also the unwritten 'organizing principles of knowledge practices' (Georgiou *et al.* 2014: 255) that are typical of the discipline. In Physics, we have a language to discuss quantum mechanics but not to discuss the difficulty of questions in assessments (Johnston *et al.* 1998). This problem becomes apparent in the compulsory internal moderation process in our department. Each assessment paper is reviewed by a staff member who is not part of the teaching team for the module. However, this process typically does not result in significant changes to assessment questions, due to the lack of a framework for judgement (Fakcharoenphol *et al.* 2015; Beutel *et al.* 2017).

The Semantics dimension of Legitimation Code Theory

LCT provides a theoretical framework for the study of knowledge practices (Maton 2014b: 2). Two aspects of LCT make it accessible and attractive to natural scientists. Firstly, it makes knowledge a key object of study so that 'the nature of what is taught and learned' (Maton 2013: 9) and disciplinary expertise are given their rightful relevance. Secondly, it offers a suite of analytical tools suitable for empirical research into social practices (Maton 2014b; Maton and Moore 2010).

The Semantics dimension is one set of analytical tools in LCT. These tools foreground knowledge practices (including words, symbols, concepts, images and any other form of expression) and involve a distinction between context-dependence (the degree to which meaning is related to context) and complexity (the degree to which meanings are condensed into a practice) (Maton 2014a; Maton and Doran 2017). In Semantics, variation in context-dependence is termed *semantic gravity* and variation in complexity is termed *semantic density*. In the coding of semantic gravity, it is conventional to code knowledge practices that are strongly embedded in, for example, everyday experience as stronger semantic gravity and more general or abstract forms of knowledge as weaker semantic gravity. The strength of semantic density is coded as stronger as more meanings are condensed into knowledge practices. The strengths of semantic gravity and semantic density may each vary along a continuum, and strengths are always relative, never absolute. It is conventional to represent these concepts on the *semantic plane*, with semantic density and semantic gravity on the two axes (Figure 3.1) Four principal modalities are identified: *rhizomatic codes* (SG–, SD+), *prosaic codes* (SG+, SD–), *rarefied codes* (SG–, SD–) and *worldly codes* (SG+, SD+).

Semantic gravity is a measure of context-dependence that is directly connected to cumulative learning and transfer. Maton (2009) argues that exposing students to knowledge practices with a wide range of semantic gravity strengths is a key condition for fostering cumulative learning. Not only should students be exposed to knowledge practices with weaker semantic gravity, but the transition between context-independent principles and

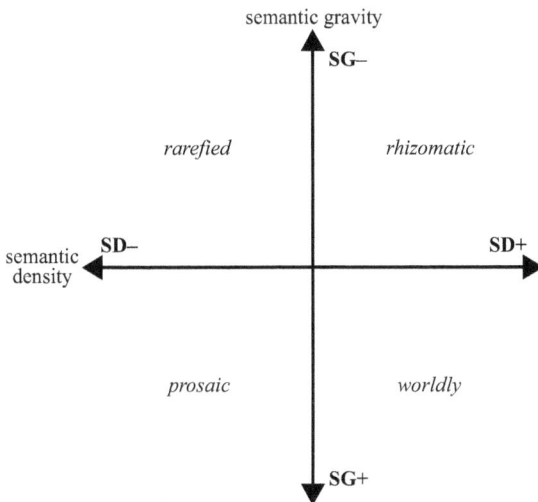

Figure 3.1 The semantic plane (Maton 2014a, 2014b: 131).

context-dependent examples, and vice versa (which are termed 'waves') should be modelled in teaching (Maton 2013). A number of papers (Kilpert and Shay 2013; Maton 2013; Conana *et al.* 2016) find that in their data, semantic gravity and semantic density are related in an inverse way: stronger semantic gravity (context-dependence) is associated with weaker semantic density (low complexity) and weaker semantic gravity (context-independence) is associated with stronger semantic density (high degree of condensation of knowledge in terms or other representations). However, Maton (2013) emphasizes that the two strengths can and do change independently so that one can encounter context-dependent but complex meanings and context-independent but simpler meanings.

An early application of LCT to assessment was done by Shay (2008) who concluded that LCT is a useful framework for conceptualizing the relation between knowledge and the criteria used in assessments. It has been concluded in several papers that cumulative learning is promoted by assessing over appropriate ranges of semantic gravity (Maton 2013; Kilpert and Shay 2013). Recently, the Semantics dimension was used to critique first-year Chemistry exam questions (Rootman-le Grange and Blackie 2018) and in our earlier work to analyze Physics assessments (Steenkamp *et al.* 2021). Semantics has also been applied to Physics education research by Georgiou *et al.* (2014) in analysis of students' responses to a thermal Physics question and by Conana *et al.* (2016) in a study of students' methods of problem-solving on the topic of mechanics in first-year Physics. In both papers, answers of students were analyzed using the wording of the question as context for exploring semantic gravity.

In our study, we apply semantic density and semantic gravity independently to analyze questions in Physics assessments. We agree that, in general, material with weaker semantic gravity tends to be expressed with stronger semantic density. However, in the context of our study objective to characterize assessment questions and answers, where a large variety of question types are included, we propose that there may be exceptions to this trend. Our approach of coding semantic gravity and semantic density independently is related to a study of Chemistry (Blackie 2014; Rootman-le Grange and Blackie 2018) but has to our knowledge not been applied before to question papers in Physics.

Background to the present study

The module on which the present study focuses is the first of two calculus-based introductory Physics modules (called Ph101 and Ph102, respectively, for the purpose of this study) that constitute the first year of a three-year Bachelor of Science programme. The modules are compulsory for students in experimental and theoretical Physics (12–20% of the class), Mathematics, Applied Mathematics, Computer Science, Earth Sciences, Geoinformatics,

Chemistry, Polymer Science and are electives for students in various biological programmes. The Ph101 curriculum introduces vector algebra and basic calculus and covers Newtonian mechanics, gravity, fluid mechanics and thermodynamics. In Ph 102 electricity, magnetism and special relativity are the main topics. 'Sears and Zemansky's University Physics' authored by Young and Freedman (2016) is the prescribed textbook. The main summative assessments in Ph101 are two tests (tests 1 and 2) and two exam opportunities for each module (exams 1 and 2).

We have previously published an analysis of test and exam papers using semantic gravity of both Ph101 and Ph102 for 2012–16 (Steenkamp *et al.* 2021). This study showed significant variation in the semantic gravity ranges that were assessed, as well as a correlation between semantic gravity and students' average marks. We concluded that the range of semantic gravity in assessments should be controlled to ensure consistent and fair assessments. The second key finding was that questions from the weakest semantic gravity category, that test knowledge of core concepts, were underrepresented in the question papers. Therefore, our assessments did not communicate the importance of the core concepts to students. This analysis led to an informed internal moderation process of new test and exam papers (referred to as the intervention) during the 2017 academic year (Steenkamp *et al.* 2021). The intervention was aimed to change our assessments so that students will be assessed consistently over appropriate ranges of semantic gravity and to emphasize the importance of core concepts, by their direct assessment. Evaluation of the intervention resulted in the hypothesis that increased focus on and assessment of core concepts lead to an improvement in the students' ability to transfer their knowledge to real-life problems on the stronger semantic gravity end of the scale. The impact of semantic density was not considered during this original study.

Aims

This chapter reports on the analysis of semantic density in the Ph101 assessments of 2016 and 2017 (before and after the intervention). The analysis was motivated by the question of whether the manipulation of semantic gravity during the intervention also influenced the semantic density of assessment questions and, if so, how this influenced student performance. The semantic density analysis was done after the papers had been finalized and written. The overarching aim of this study is to determine what we can learn by using Semantics to analyze assessment questions in first-year Physics modules and whether such analysis is useful for making changes in educational practices.

By combining the current semantic density analysis with previously produced semantic gravity analysis data, we aimed to answer the following questions, using the intervention during 2017 as a test case:

- What does analysis of the context-dependence and complexity of assessment questions reveal about what we really require of students?
- Is there a correlation between the semantic density and semantic gravity strengths of questions or answers and students' ability to answer the questions (marks)?
- Did the intervention based on semantic gravity also unintentionally change the semantic density of the assessments?
- Would semantic density analysis support or falsify the hypothesis that the increased focus on core concepts during the intervention improved students' ability to transfer their knowledge to problems with stronger semantic gravity?

The current study is focused on the assessments of Ph101 as the intervention had the largest effect on this module. On the longer time scale, we aim to continue using Semantics for self-critique of our assessments in order to facilitate informed change to teaching and assessment practices.

Methodology

LCT concepts are highly abstract and general, in order to be applicable to a wide range of different practices. Thus, for each study one needs a 'translation device' (Maton 2014b; Maton and Chen 2016) that shows how a concept is realized in the particular data being analyzed. The development of a 'translation device' for enacting semantic density is discussed below. The semantic gravity analysis that we used has been published before, and thus the translation device is only discussed briefly (Steenkamp *et al.* 2021). Both devices were developed in order to be useful for all types of questions in our assessments.

The translation device for enacting semantic density, as shown in Table 3.1, considers the different representations of knowledge that are generally used in Physics to condense meaning. The use of representations is an active field in Physics education research (Van Heuvelen 1991; Fredlund *et al.* 2014; Docktor and Mestre 2014). In their paper pioneering the application of Semantics in Physics, Conana *et al.* (2016) also referred to different types of representations in their analysis of students' problem-solving approaches. The development of our translation device was guided by recent work by Maton and Doran (2017) on English discourse, which suggests that the more meanings are condensed within a representation, the stronger the semantic density. The typical representations used in Physics include verbal descriptions complemented by sketches and diagrams, more specialized vector diagrams or graphs and mathematical representations.

From weaker towards stronger semantic density, our translation device differentiates: verbal descriptions and images providing no technical information (SD−−), descriptions and images that does provide technical

Table 3.1 Translation device for semantic density

Category	Criteria
SD++	A mathematical representation that has been manipulated and condensed to construct a mathematical model for the problem.
SD+	A mathematical representation expressing relations between key concepts.
SD0	A graph conveying relationships between key concepts or a vector diagram conveying the relationships between the vector quantities in the problem.
SD–	Verbal representation of the problem using the key technical terms or symbols and/or a sketch or diagram conveying technical details.
SD––	Verbal representation of the problem in everyday words and/or a photo or sketch conveying the important features of the problem visually but using no technical terms and conveying no technical details.

information that links them to core concepts (SD–) and graphs or vector diagrams (SD0) that provide information regarding relations of concepts that is more detailed than can be given in verbal descriptions. Mathematical representations consist of symbols representing physical quantities and occasionally contain numerical values. We consider a mathematical representation that expresses the relation between two or more key concepts to be typically of higher semantic density (SD+) than a graph of the same relation (SD0) as the mathematical expression usually contains additional meaning. Furthermore, the 'SD+' category simply requires core concepts and their basic relations to have been written down as mathematical expressions, while in the 'SD++' category mathematical expressions are manipulated to construct a mathematical model for a specific problem. The model then contains more meaning than the collection of expressions typical of the 'SD+' category, because relations that were not apparent in the 'SD+' category are now made explicit and typically the relation between parameters becomes more complex.

In the coding of semantic density, we disregard the numerical step that is sometimes required in assessments – namely, to substitute numbers into the mathematical expression and subsequently perform numerical calculations in order to obtain a numerical answer. In our assessments, this step usually counts very little in terms of mark allocation. Table 3.2 contains some examples of our coding of the various assessment questions, to further clarify the application of the translation device.

In the semantic density analysis, the assessment questions and model answers were coded independently. In compound questions the semantic density of each numbered sub-section was coded separately. When a question or an answer included elements of different semantic density categories,

Table 3.2 Examples of the coding of assessment questions representing different categories of semantic density and semantic gravity, along with an explanation to justify the allocated coding

Question	Model answer	Explanation of coding
Explain briefly what $\hat{j}$ represents in the context of vectors. Would it be correct to say that $\hat{j}$ is the same as 1? Explain your answer.	$\hat{j}$ represents the unit vector in the positive y direction. $\hat{j}$ is a vector with a direction and is therefore not the same as the scalar 1.	The semantic gravity is coded SG-- as there is no reference to any empirical example. The semantic density is relatively weak (SD–) for both question and answer as both are verbal descriptions including symbols with technical meaning.
A bat strikes a 0.145 kg cricket ball. Just before the impact, the ball is travelling horizontally to the right at 60.0 m/s. After being struck by the bat, the ball travels to the left at an angle of 35 degrees above the horizontal with a speed of 60.0 m/s. the ball and bat are in contact for 1.85 ms. Determine the horizontal and vertical components of the average force on the ball.	Write the momentum of the ball before and after being hit as vectors: $$\vec{p}_1 = (60)\hat{i}$$ $$\vec{p}_2 = -(60)\cos(35\,degrees)\hat{i} + (60)\sin(35\,degrees)\hat{j}$$ Change in momentum is related to average force via the impulse. $$\vec{J} = \vec{p}_2 - \vec{p}_1 = \vec{F}\Delta t$$ $$\vec{F} = \frac{\vec{p}_2 - \vec{p}_1}{\Delta t}$$ Substitute the numbers and calculate final vector.	The semantic gravity is relatively weak (SG–) as one core concept – namely, impulse – is applied to a simplified scenario closely associated with that concept. The semantic density of the question is coded as SD-- as it is a verbal description in everyday words. The semantic density of the answer is stronger (SD+) since the expected answer requires the use of mathematical expressions.

| At a time $t = 0$ s a particle 1 has the position $r_1 = 30\hat{j}$ m. It always moves at a constant velocity $v_1 = 3.0\hat{i}$ m/s. Another particle, particle 2, has zero velocity at time $t = 0$ s, and it has a constant acceleration $\vec{a}$, with magnitude $a = 0.40$ m/s². Particle 2 is at the origin at $t = 0$ s. There is no gravity. Calculate at which angle θ the acceleration vector $\vec{a}$ must point with respect to the positive y-axis so that the two particles v collide.

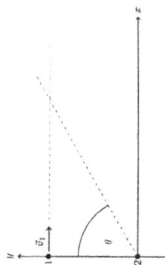 | The student must apply this argument: we need to compute the trajectories of both particles. They will collide if the x and y components of the two trajectories are identical at the same time. Particle 1 has a constant velocity:

$$\vec{r_1}(t) = (3.0)t\hat{i} + (30)\hat{j}$$

Particle 2 starts at origin with zero velocity but constant acceleration. The magnitude of its displacement is

$$r_2(t) = \frac{1}{2}(0.4)t^2\hat{i} + (30)\hat{j}$$

Written as vector

$$\vec{r_2}(t) = \frac{1}{2}(0.4)\sin\theta\, t^2\hat{i} + \frac{1}{2}(0.4)\cos\theta\, t^2\hat{j}$$

Set x and y components equal at the time of collision:

$$x: (3.0)t = \frac{1}{2}(0.4)\sin\theta\, t^2$$
$$y: 30 = \frac{1}{2}(0.4)\cos\theta\, t^2$$

Combine these relations and solve simultaneously (a fair amount of work still) to obtain $\theta = 60$ degrees. | The semantic gravity is relatively strong (SG++) as this problem requires the student to construct a self-defined logical argument, involving the translation of everyday knowledge of a collision into a mathematical condition. The question contains a verbal description with technical terms (SD–) and a graph (SD0). The answer has stronger semantic density (SD++) as it requires significant manipulation and condensation of mathematical relations. |
| Water has the important property that water has the highest density at 4 degrees Celsius. Water at higher or lower temperatures, and ice, have lower densities. Explain why this property is important to prevent large water bodies (like lakes) from freezing solid down to the bottom. | The answer requires a verbal explanation, including discussion that this property of water prevents convection in water at temperatures lower than 4 degrees Celsius, thus freezing from the top and that a surface layer of ice acts as thermal insulator. | The semantic gravity is relatively strong (SG++) as the student must combine the given information and core principles, combined with everyday knowledge, to a specific 'real-world' scenario. The semantic density of both the question and answer is SD–, as both consist of verbal representations using technical terms. |

we reported the strongest semantic density. The motivation for this decision is that every question or answer will include text of weaker semantic density, and it is of little meaning to code that if elements of stronger semantic density are present.

The translation device for semantic gravity has been published before, and a more detailed discussion is given in the original publication (Steenkamp *et al.* 2021). The translation device is reproduced in Table 3.3. For this translation device, four categories were defined, labelled: SG--, SG-, SG+ and SG++. The weakest category, 'SG--' , is associated with formulating a core concept in the general context-independent form, 'SG-' is the application of a concept to a simplified and idealized scenario, 'SG+' is the application to a well-defined empirical scenario and 'SG++' represents the application of core concepts to a real-world problem, where self-defined assumptions and application of everyday knowledge is required in addition to the concepts. In the case of semantic gravity, the question with its model answer is coded as a unit, contrary to the independent coding of question and answer in semantic density. Most assessment questions include aspects of more than one semantic gravity category. For the purpose of combining our semantic gravity and semantic density analyses, we reported the strongest semantic gravity categories found in those questions. Examples of how the translation device was applied to our data are presented in Table 3.2.

The average marks of the cohort of students for individual questions, or sub-questions, were used as a measure of the students' ability to master the semantic gravity and semantic density present in the question and required for a successful answer. In order to compare marks before (2016) and after the intervention (2017) the group of first-time enrolled students were selected, and the marks of students who were repeating the module were not considered. The data also excluded the small number of students who deregistered or changed to a different programme during the academic year.

Table 3.3 Translation device used for semantic gravity levels.

Level	Criteria
SG++	Application of a core concept to a 'real-world' problem that can only be solved by additionally applying everyday knowledge and self-defined assumptions.
SG+	Application of a core concept to a well-defined empirical scenario, where the association of the core concept with the scenario has not been discussed in the curriculum, although the scenario lies within the scope of the curriculum.
SG-	Application of a core concept to a simplified empirical scenario that is associated with this specific core concept in the curriculum.
SG--	Formulation of a core concept (a general principle, concept, definition or law) that is found in one clearly defined section of the curriculum, without reference to an example from the empirical domain.

From Steenkamp *et al.* (2021), reprinted by permission of the publisher (Taylor & Francis Ltd, http://www.tandfonline.com)

For a detailed comparison, the marks to the first exam of Ph101 were analyzed on the level of sub-questions. This was done for 50% of the cohort, selecting every second name on an alphabetical list as a representative sample.

Results and discussion

Figure 3.2 shows the semantic density analysis of the assessments of 2016 and 2017. The graphs show the percentage of marks allocated to either questions (a) or answers (b) for each of the four semantic density categories. We will call these the semantic density 'maps' of the assessment questions or answers.

It is observed that the questions generally show weaker semantic density than the model answers. The semantic density of the questions include a few questions in the 'SD--' category (non-technical verbal descriptions), a large majority of questions in the 'SD-' category (verbal descriptions involving technical terms), the use of graphs and vector diagrams in the 'SD0' category and of mathematical expressions in the 'SD+' and 'SD++' categories. The model answers generally exclude non-technical verbal descriptions (SD--). Technical verbal descriptions (SD-) comprise on average 10%–20% of the answers. Graphs and diagrams (SD0) form a small fraction of answers, writing down basic mathematical expressions without significant manipulation (SD+) makes up 20–30% of answers. More than 50% of answers require the construction of mathematical models by manipulating and condensing the basic relations to derive additional relations (SD++).

It is evident that the intervention caused observable changes in the semantic density maps. In both the questions and the answers, the semantic density weakened on average. In both questions and answers, the use of verbal descriptions using technical terms (SD-) increased while the use of mathematical expressions (SD+) and models (SD++) decreased during 2017 in comparison to 2016. This reflects a larger focus on core concepts and a

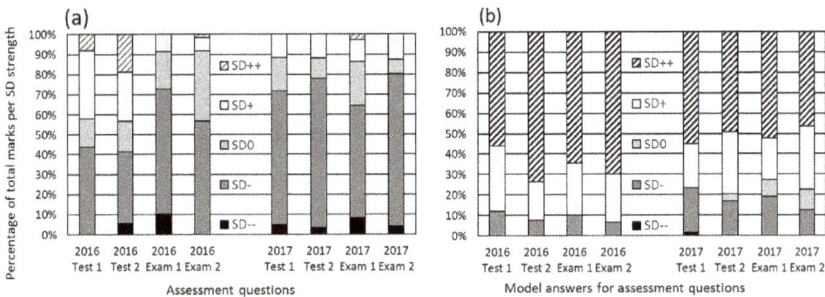

Figure 3.2 Semantic density maps of the Ph101 assessments of 2016 and 2017 representing the percentage of marks allocated to (a) questions and (b) model answers of different semantic density categories.

realization that assessment of conceptual understanding should include verbal explanations. Compared to the semantic gravity maps of the same assessments (see Figure 3.4(a) of Steenkamp *et al.* 2021), the change in the semantic density map is less pronounced than the change in the semantic gravity map. This is probably because the semantic gravity map was modified on purpose, whereas the change in semantic density was unintentional.

The use of mathematical expressions in questions during 2016 and 2017 differs in an interesting way. In both years, some questions contain mathematics that form an integral part of the question, but questions also occur where non-essential mathematical expressions are given. These questions can be answered without the mathematical expression, but the expression serves as 'scaffolding' (Dawkins *et al.* 2017). Considering these questions, the expressions given during 2016 were often the expression for the final answer that must be derived or proven (coded as SD++). This communicates an emphasis on reaching a certain answer. During 2017, the given expressions were usually representing one of the core concepts that serve as a starting point for answering the question (coded as SD+), communicating an emphasis on core concepts. This may be the result of the lecturers' increased awareness of core concepts.

Graphs and vector diagrams (SD0) were used in questions in both years but were present in answers only during 2017. In these 2017 questions, conceptual understanding was tested by requiring students to sketch a graph or a vector diagram.

We found it useful to characterize assessment question-answer pairs by plotting these as points on a graph that has the semantic density of the question on the horizontal axis and the semantic density of the answer on the vertical axis (Figure 3.3). The areas of the circles in Figure 3.3 represent the number of question-answer pairs in each category. What this analysis revealed is that most of the data points are above the upwards sloping diagonal, meaning that the model answers generally have stronger semantic density than their associated questions. However, there are questions in each test positioned below the

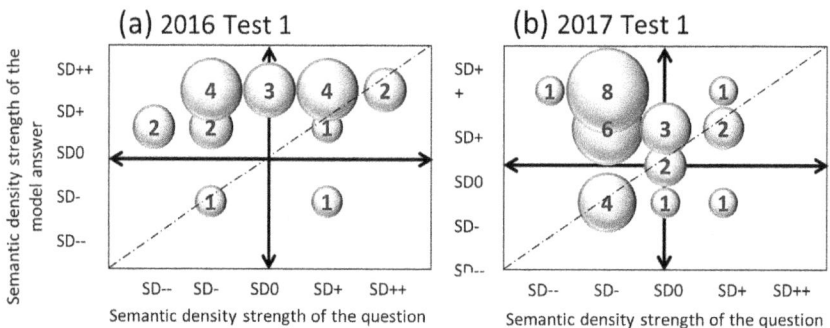

Figure 3.3 Number of question-answer pairs plotted on a two-dimensional graph representing the semantic density of answers versus questions, for the main tests of 2016 (a) and 2017 (b), respectively.

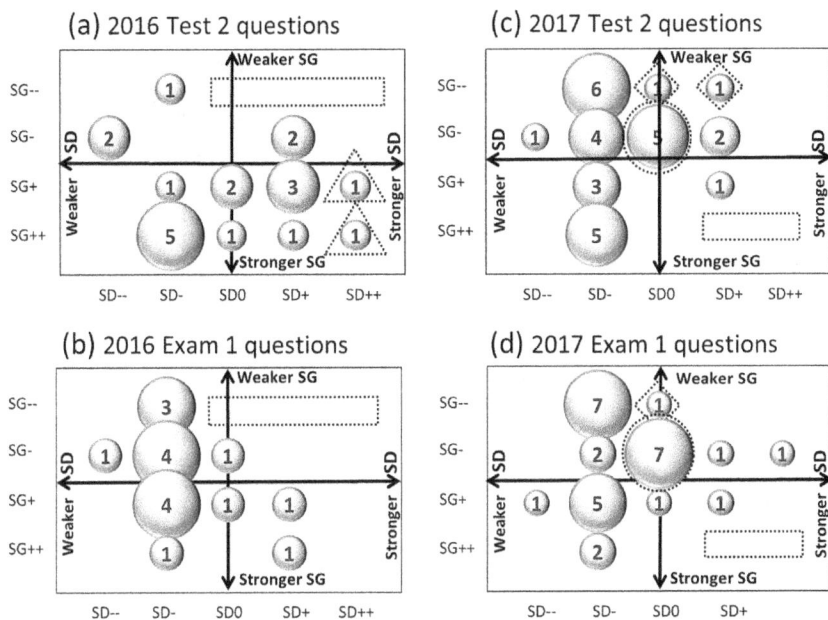

Figure 3.4 The questions of the class tests of 2016 (a) and 2017 (c) and first exams of 2016 (b) and 2017 (d) of Ph101, plotted on the semantic plane.

diagonal, where a verbal description (of weaker semantic density) was required as answer for a question containing a diagram or mathematics (of stronger semantic density). In the 2016 test, the model answers were mostly in the form of mathematical expressions (SD+ and SD++ on the vertical axis), whereas a wider distribution of answer types is seen in the 2017 test. We conclude that the purposeful changes to the semantic gravity maps of the question papers during 2017 did influence the semantic density of both the questions and required answers.

The semantic plane offers a visualization of the relation between semantic gravity and semantic density. Figure 3.4 shows the semantic planes for the questions from four analyzed assessments, while Figure 3.5 shows the semantic planes for the model answers of these assessments. In both figures, the areas of the circles and numbers inside represent the number of question-answer pairs in each category. Dotted lines are guides for the eye and are referred to in the discussion.

Figure 3.4 shows that the questions cover the full range of semantic gravity and semantic density, with the SD– category the most prominent in terms of semantic density. In the 2016 papers (also typical for the other 2016 assessment not shown here), there is a trend that a stronger semantic density is associated with a stronger semantic gravity and vice versa, seen by the dominant grouping of questions in the rarefied and worldly codes, in comparison to the rhizomatic and prosaic codes. Furthermore, questions coded 'SG--,' that directly assess core concepts, are asked as technical verbal

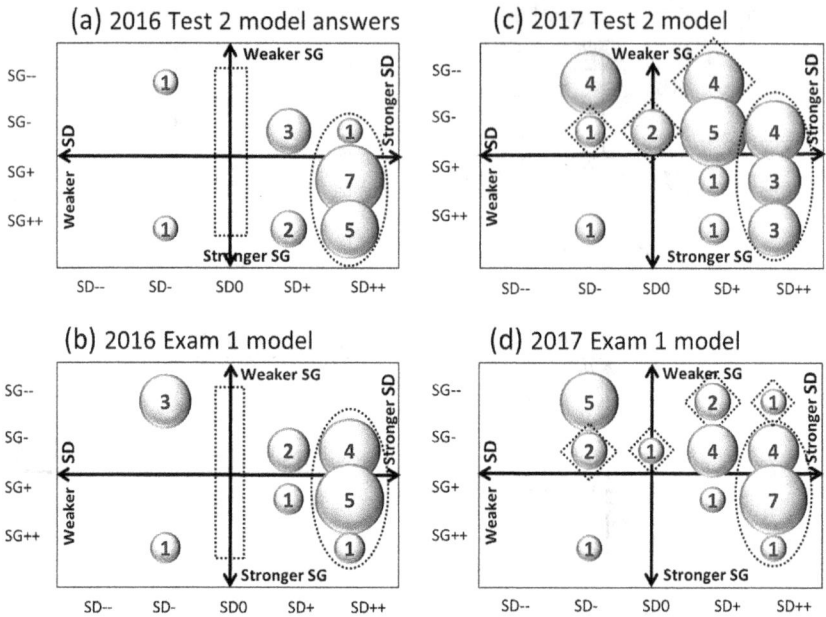

Figure 3.5 The model answers of the class tests of 2016 (a) and 2017 (c) and first exams of 2016 (b) and 2017 (d) of Ph101, plotted on the semantic plane.

descriptions (SD–) only, and questions of semantic gravity SG–– that have stronger semantic densities (indicated by dotted rectangles in graphs (a) and (b)) are typically lacking. Questions in worldly codes (dotted triangles) with both stronger semantic gravity (SG+ and SG++) and stronger semantic density (SD++) were frequently asked before the intervention. These trends differ from the association of weaker semantic gravity with stronger semantic density used in other studies (Conana *et al.* 2016). The reason for this is that before the intervention, typically questions that required applications to real-life scenarios (SG+ or SG++) had to be solved by constructing a mathematical model, and some of that mathematics was given in the question as scaffolding (SD+ or SD++), whereas questions on core concepts (SG–– or SG–) were mostly assessed using verbal descriptions (SD–– or SD–).

In 2017, this trend was absent. The semantic gravity category 'SG––,' where core concepts are assessed directly, was assessed in more diverse ways, using questions with stronger semantic density (dotted diamond shapes). The number of questions where a graph or vector diagram is given in the question and assessed within an intermediate semantic gravity range (dotted circles) was increased. This resulted in a distribution of questions between the rarefied, rhizomatic and prosaic codes.

The lack of questions in the worldly code representing stronger semantic gravity and stronger semantic density – rectangles in graphs (c) and (d)) in the 2017 assessments – may be flagged. At first impression, it may be asked whether the 'hardest' questions have been omitted from the 2017 assessments. However, closer analysis revealed that in the two SD++ questions in the 2016 test, the mathematical expression of the final answer is given, and the student is asked to prove that this is true. The same Physics has been assessed in the 2017 test without giving the final answer as part of the question, thus weakening the semantic density of the question without changing the Physics that is assessed.

A meaningful use of the SD++ question category would be to give a mathematical model of a real-life scenario and ask students to interpret the model and draw conclusions. This example illustrates how plotting the questions on the semantic plane can be useful for critical self-evaluation of a question paper by giving an overview of the types of questions, but that not all questions in a particular SG and SD category are equal in terms of difficulty.

Figure 3.5 shows the coding of the model answers on the semantic plane. A clear difference between 2016 and 2017 assessments can be seen. In the 2016 assessments, most of the answers are clustered in the worldly code (dotted oval in Figure 3.5(a)) representative of stronger semantic density (SD++) and intermediate to stronger semantic gravity categories (SG–, SG+, SG++). Furthermore, students were not asked to draw graphs or vector diagrams (SD0) as answers (dotted rectangle). Answers in the 'SG––' category are only in the form of technical verbal descriptions (SD–). It thus shows that the intervention caused us to set questions that require answers over a wider variety of semantic gravity and semantic density categories. New types of questions (indicated by dotted diamond shapes) include answers on core concepts (SG––) with stronger semantic density (SD+) and applications of core concepts (SG–) that require a graph (SD0) or verbal description (SD–). Answers requiring the construction of a mathematical model (SD++) remain important, as indicated by the dotted ovals in Figure 3.5(c) and (d).

The results confirm changes in the semantic density, as well as semantic gravity, due to the intervention. In 2017, the questions and answers are distributed more evenly over a wider range of semantic gravity and semantic density categories. An increased focus on weaker semantic gravity (SG––) is also observed. In terms of semantic density, technical verbal representations as questions and mathematical expressions as answers remain dominant, corresponding to the semantic density maps in Figure 3.2.

An important question remains to be answered: do student marks correlate with question-answer pairs of different semantic gravity and semantic density codes? To investigate this, we performed a detailed analysis of the average marks of the first exams of 2016 and 2017 on the level of subquestions in order to calculate the average mark for questions of a particular semantic gravity category and semantic density category. We selected exam 1 since it is the most important assessment. Results from the 2016 and 2017 papers were combined.

We have found that the most useful way to study the effect on marks is to visualize the dependence of marks on semantic gravity and semantic density as in Figure 3.6. Figure 3.6(a) and (c) show the variation of marks over the different semantic density categories of the questions and answers,

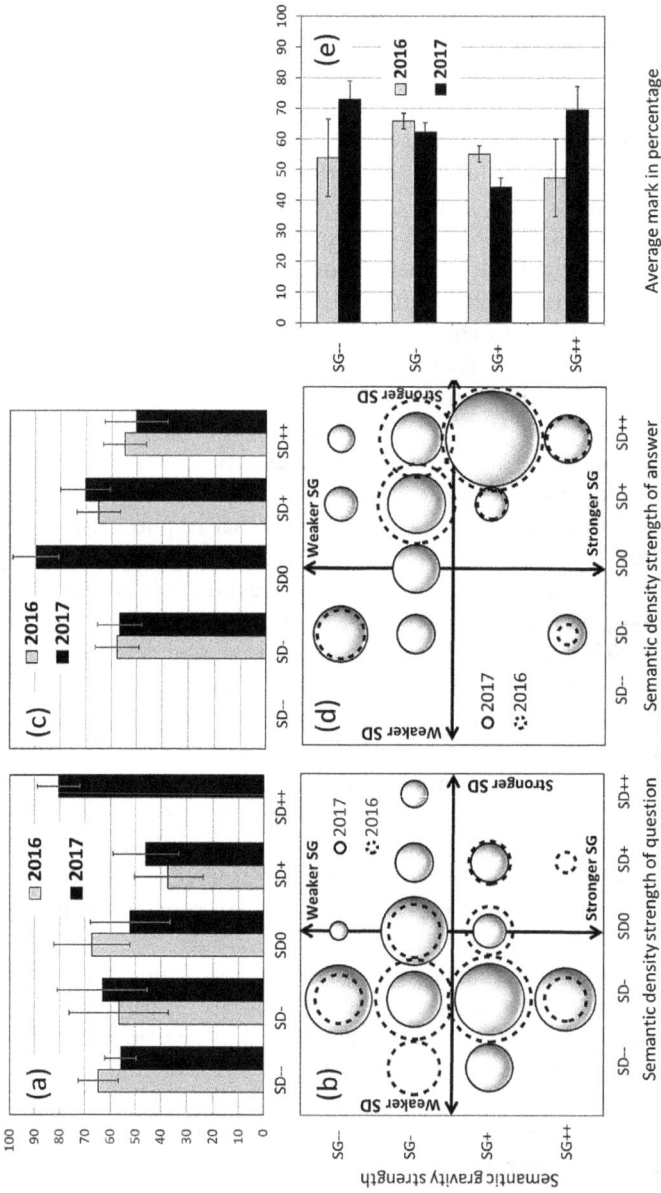

Figure 3.6 The average marks plotted versus the semantic density of the question (a) and answer (c) and the semantic gravity (e), respectively, shown with the weights of the different categories of questions (b) and answers (d) plotted on the semantic plane.

respectively, whereas Figure 3.6(e) shows the variation of marks over the different semantic gravity categories for question-answer-pairs combined. The error bar lengths in these graphs agree to the standard deviations of the marks. To facilitate interpretation of these graphs, they are displayed with plots of the weights of the question-and-answer categories on the semantic plane – Figure 3.6(b) and (d). In these plots, the diameter of each circle represents the number of marks allocated to the category as percentage of the total of the exam paper. It means that, for example, all the questions represented in the vertical SD+ column of graph (b) contribute to the marks represented by the SD+ bar in graph (a), and all the questions represented in the horizontal SG– row of graph (b) contribute to the marks in the SG– bar in graph (e).

The variation of marks over the semantic gravity categories (Figure 3.6(e)) shows different trends during 2016 and 2017. During 2016, the marks were highest for intermediate semantic gravity categories (SG– and SG+) and students did not perform as well as expected in the weakest category (SG––). This means that students struggled to identify the correct core concept or to formulate or explain the concept correctly. Marks are also lower, as expected, when the semantic gravity is stronger (SG++), and a significant degree of transfer of knowledge to a real-life scenario is required. During 2017, this trend became inverted so that the weakest category (SG––) correlated with the highest average mark. This confirms that the efforts to encourage students to know the core concepts well have been successful. The surprising result was that the marks in the strongest category (SG++) also increased significantly. The question was asked whether these differences could be caused by unintended changes in semantic density. This question can be investigated using Figure 3.6, as the plot on the semantic plane links the graph of marks versus semantic gravity to the graph of marks versus the semantic density.

In the weakest semantic gravity category (SG––) in Figure 3.6(b), the questions were of approximately the same semantic density (mostly SD–) during 2016 and 2017. The answers (Figure 3.6(d)) were extended to stronger semantic density (SD+ and SD++) during 2017. It means that the 2017 students performed better in questions testing core concepts, although answers of stronger semantic density were expected in this category. In the strongest semantic gravity category (SG++), the semantic density of questions (Figure 3.6(b)) and answers (Figure 3.6(d)) during 2016 and 2017 were similar with a slightly higher weight given to weaker semantic density questions and answers in 2017, meaning that the improved performance of students in the SG++ category during 2017 cannot be the result of a significant change in semantic density. This supports the hypothesis that the higher marks in the weaker and stronger semantic gravity categories (SG–– and SG++) are indeed linked: the increased awareness of what the core concepts are and being able to formulate them correctly enables students to transfer their knowledge to real-life scenarios.

In SG– and SG+ categories, the marks are lower during 2017 than during 2016, but the decreases are relatively small (4% and 11% lower in the

SG– and SG+ categories, respectively). In the SG– category, the semantic density of the questions was extended towards stronger semantic density and the answers towards weaker semantic density during 2017. In the SG+ category, the questions were extended towards weaker semantic density and the semantic density of the answers is unchanged from 2016 to 2017. This means that the marks of the SG– and SG+ categories show a similar decrease from 2016 to 2017 in spite of opposing changes to the semantic density from 2016 to 2017.

The general conclusion is therefore that semantic density is not clearly correlated with the changes in marks from 2016 to 2017 in either of the four semantic gravity categories. Considering the variation of marks with the semantic density of the question (Figure 3.6(a)) a significant difference between the 2016 and 2017 marks (difference larger than the error bar representing the standard deviation) exists only in the SD–– category. Considering the data of both 2016 and 2017 together (and ignoring the SD++ bar of 2017 as it represents a single low-weight question only), the general trend is a decrease in marks as the semantic density of the question is increased. In this trend, the semantic gravity should not play a large role as the mapping of the questions on the semantic plane (Figure 3.6(b)) is approximately symmetric around the horizontal axis so that the trend must be caused by the increase in semantic density. The variation of marks with the semantic density of the answers (Figure 3.6(c)) shows a pronounced decrease of marks from SD0 to SD++. We consider this to be caused by both the combination of the increasing semantic gravity and increasing semantic density of the required answers.

We thus found evidence of an influence of semantic density on marks, but this effect is weaker than that of the semantic gravity. It thus seems that our students are more successful in working with different representations of knowledge than to transfer knowledge to unfamiliar contexts. The influence of increasing semantic density on marks is most pronounced in the SD0, SD+ and SD++ where the increasing semantic density is associated with graphs and symbolic mathematical representations. This correlates with observational evidence that many first-year students struggle to manipulate and interpret symbolic mathematical expressions and graphs.

The final question regarding the influence of semantic density on marks is whether it may be the difference between the semantic density of the question and the semantic density of the answer that plays a role, rather than any one of these separately. This is investigated by plotting the average marks versus the semantic density difference (SDD) in Figure 3.7, where the sign and number indicate the measure of strengthening (+) or weakening (–) of the semantic density from question to answer. The error bar length agrees to the standard deviation of the marks. For example, SDD+1 represents a question-answer pair for which the semantic density of the answer is one category stronger than that of the question. In this graph, the 2016 and 2017 marks show opposing trends. When ignoring the low number of data points in the SDD–2 and SDD–1 categories, the 2017 marks decrease

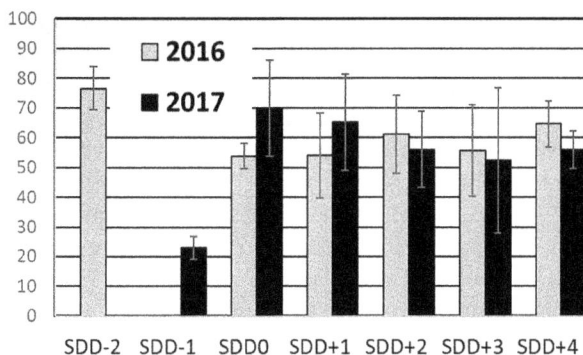

Figure 3.7 Average marks in the first exams of 2016 and 2017 as a function of the SDD.

towards larger SDDs, whereas the 2016 marks show an increasing trend over this range. We conclude that the SDD on its own cannot be a critical quantity.

The analysis including both semantic density and semantic gravity confirmed the conclusion previously proposed on the basis of semantic gravity only (Steenkamp *et al.* 2021). The intervention caused a significant increase in marks in both the weaker and the stronger ends of the semantic gravity scale used. We propose that the focus on core concepts in teaching and assessment caused by the intervention communicated the importance of core concepts clearly to students, encouraging them to study these. Improved familiarity with the core concepts and their correct formulation enabled students to transfer their knowledge to problems linked to real-life scenarios. As the core concepts in Physics are formulated to be highly context-independent (abstract), this result corresponds with the idea that significant exposure of learners to knowledge of weaker semantic gravity is not only supportive of but also a condition for cumulative learning (Maton 2009; Maton 2013).

Conclusion

We have used the Semantics dimension of LCT to evaluate the outcomes of an intervention in an introductory Physics module. The intervention was the result of a previous study, which employed semantic gravity to analyze the context-dependence of the assessment questions in historical papers and resulted in an internal moderation process during which the results were used to evaluate and modify new papers (Steenkamp *et al.* 2021). We conclude that the analysis of assessments using Semantics is a practical starting point for making changes in introductory Physics modules. We argue that assessment has a direct effect on learning as it is a high-stakes activity for students and lecturers, and it communicates the expected outcomes of learning concretely to students. Our study of question papers before and during the intervention (2016 and 2017) showed the value of both semantic gravity and semantic density in categorizing types of questions and model answers in assessments. Correlation of the question categories with students' average marks has confirmed that semantic gravity has the dominant effect on marks. We have observed a trend of decreasing average marks with increasing semantic density of either the question or the answer, but the effect is weaker than that of the semantic gravity.

The analysis including both semantic density and semantic gravity supported the hypothesis that an increased focus on core concepts (weaker semantic gravity) in teaching and assessment improved the ability of our students to transfer their knowledge to questions related to real-life scenarios (stronger semantic gravity). This is a step towards cumulative learning.

We conclude that the Semantics dimension of LCT is a valuable analytical tool to Physics lecturers and Physics education specialists. LCT acknowledges that disciplinary knowledge has a unique character and aims to study this in a systematic way, thus avoiding a 'knowledge-blind' approach to education (Maton 2013). We agree with the conclusions of Georgiou *et al.* (2014) that in Physics education the forms that knowledge takes play an important role in learning and that LCT, therefore, offers a productive approach to analyze educational practice and, in particular, evaluate changes in educational practice in the interest of informed decisions. LCT is, however, more than a tool, but a theoretical framework for the study of knowledge and useful to guide changes to many aspects of education.

Acknowledgements

The authors wish to thank Zhigang Yu, Tongji University, for helpful discussions about the semantic density of images. We wish to thank Margaret Blackie for valuable advice. Christine M. Steenkamp acknowledges the support from colleagues Kristian Müller-Nedebock, Gurthwin Bosman and Hannes Kriel.

References

Barnett, S. and Ceci, S.J. (2002) 'When and where do we apply what we learn? A taxonomy for far transfer,' *Psychological Bulletin* 128(4): 612–637.

Bearman, M., Dawson, P., Boud, D., Bennett, S., Hall, M. and Molloy, E. (2016) 'Support for assessment practice: Developing the assessment design decisions framework,' *Teaching in Higher Education* 21(5): 545–556.

Bernstein, B. and Solomon, J. (1999) '"Pedagogy, identity and the construction of a theory of symbolic control" Basil Bernstein questioned by Joseph Solomon,' *British Journal of Sociology of Education* 20(2): 265–279.

Beutel, D., Adie, L. and Lloyd, M. (2017) 'Assessment moderation in an Australian context: Processes, practices, and challenges,' *Teaching in Higher Education* 22(1): 1–14.

Blackie, M.A.L. (2014) 'Creating semantic waves: Using Legitimation Code Theory as a tool to aid the teaching of Chemistry,' *Chemistry Education Research and Practice* 15(4): 462–469.

Brown, S. and Knight, P. (1994) *Assessing learners in higher education*, Kogan Page.

Conana, H., Marshall, D., & Case, J. M. (2016). Exploring pedagogical possibilities for transformative approaches to academic literacies in undergraduate Physics. *Critical Studies in Teaching and Learning*, 4(2), 28–44.

Dawkins, H., Hedgeland, H. and Jordan, S. (2017) 'Impact of scaffolding and question structure on the gender gap,' *Physical Review Physics Education Research* 13(2): 020117.

Docktor, J.L., and Mestre, J.P. (2014) 'Synthesis of discipline-based education research in Physics,' *Physical Review Physics Education Research* 10(2): 020119.

Fakcharoenphol, W., Morphew, J.W. and Mestre, J.P. (2015) 'Judgements of physics problem difficulty among experts and novices,' *Physical Review Physics Education Research* 11(2): 020128.

Finkelstein, N. (2005) 'Learning physics in context: A study of student learning about electricity and magnetism,' *International Journal of Science Education* 27(10): 1187–1209.

Fredlund, T., Linder, C., Airey, J. and Linder, A. (2014) 'Unpacking physics representations: Towards an appreciation of disciplinary affordance,' *Physics Education Research* 10(2): 020129.

Georgiou, H., Maton, K. and Sharma, M. (2014) 'Recovering knowledge for science education research: Exploring the "Icarus effect" in student work,' *Canadian Journal of Science, Mathematics and Technology Education* 14(3): 252–268.

Johnston, I.D., Crawford, K. and Fletcher, P.R. (1998) 'Student difficulties in learning quantum mechanics,' *International Journal for Science Education* 20(4): 427–446.

Kilpert, L. and Shay, S. (2013) 'Kindling fires: Examining the potential for cumulative learning in a Journalism curriculum,' *Teaching in Higher Education* 18(1): 40–52.

Laverty, J.T., Underwood, S.M., Matz, R.L., Posey, L.A., Carmel, J.H., Caballero, M.D., Fata-Hartley, C.L., Ebert-May, D., Jardeleza, S.E. and Cooper, M.M. (2016) 'Characterizing college science assessments: The three-dimensional learning assessment protocol,' *PLoS One* 11(9): e0162333.

Lobato, J. (2006) 'Alternative perspectives on the transfer of learning: History, issues, and challenges for future research,' *The Journal of the Learning Sciences* 15(4): 431–449.

Maton, K. (2009) 'Cumulative and segmented learning: Exploring the role of curriculum structures in knowledge-building.' *British Journal of Sociology of Education* 30(1): 43–57.

Maton, K. (2013) 'Making semantic waves: A key to cumulative knowledge building,' *Linguistics and Education* 24(1): 8–22.

Maton, K. (2014a) 'A TALL order? Legitimation Code Theory for academic language and learning,' *Journal of Academic Language and Learning* 8(3): A34–A48.

Maton K. (2014b) *Knowledge and Knowers: Towards a Realist Sociology of Education*, Routledge.

Maton, K. and Chen, R. T-H. (2016) 'LCT in qualitative research: Creating a translation device for studying constructivist pedagogy,' in K. Maton, S. Hood and S. Shay (eds) *Knowledge-Building: Educational Studies in Legitimation Code Theory*. Routledge, 27–48.

Maton, K. and Doran, Y.J. (2017) 'Semantic density: A translation device for revealing complexity of knowledge practices in discourse, part 1—wording,' *Onomázein* (1): 46–76.

Maton, K. and Moore, R., eds. (2010) *Social realism, knowledge and the sociology of education: Coalitions of the mind.* Continuum.

Nokes-Malach T.J. and Mestre, J.P. (2013) 'Toward a model of transfer as sense-making,' *Educational Psychologist* 48(3): 184–207.

Rootman-le Grange, I. and Blackie, M.A.L. (2018) 'Assessing assessment: In pursuit of meaningful learning,' *Chemistry Education Research and Practice* 19(2): 484–490.

Shay, S. (2008) 'Beyond social constructivist perspectives on assessment: The centring of knowledge,' *Teaching in Higher Education* 13(5): 595–605.

Steenkamp, C. M., Rootman-le Grange, I., & Müller-Nedebock, K. K. (2021). Analysing assessments in introductory physics using semantic gravity: Refocussing on core concepts and context-dependence. *Teaching in Higher Education, 26*(6), 871–886.

Van Heuvelen, A. (1991) 'Learning to think like a physicist: A review of research-based instructional strategies,' *American Journal of Physics* 59: 891–897.

Walsh, L.N., Howard, R.G. and Bowe, B. (2007) 'Phenomenographic study of students' problem solving approaches in Physics,' *Physics Education Research* 3(2): 020108.

Young, H.D. and Freedman, R.A. (2016) *Sears and Zemansky's University physics with modern physics* (14th ed.). Pearson Education.

4 Building complexity in Chemistry through images

Zhigang Yu, Karl Maton, and Yaegan Doran

Introduction

Images are common in Chemistry. Photographs, diagrams, graphs and charts are widely used to represent Chemistry knowledge and form a crucial component of the texts through which students learn that knowledge. A key feature of images is the complexity of meanings they express, as different degrees of complexity are needed in different learning stages (Dimopoulos *et al.* 2003, Kapıcı and Savaşcı-Açıkalın 2015, Pintó and Ametller 2002). For example, Figure 4.1 includes two images from Chemistry textbooks designed for secondary school curriculum in New South Wales (NSW), Australia.

The image on the top is from a Year 7 textbook discussion of states of matter and shows an 'everyday' phenomenon: ice melting to become liquid water. The epistemological meanings expressed are relatively simple. In contrast, the image on the bottom is from a Year 11 textbook and illustrates the working mechanism of a 'Daniell cell,' a type of electrochemical cell that converts chemical energy into electrical energy. The diagram presents multiple technical elements and processes that are key to the energy conversion. The complexity of the diagram can be illustrated by unpacking the meanings being expressed into a written description. The experimental set-up includes three main components: cathode, anode and salt bridge. The cathode is 'copper' in the right beaker and shown gaining copper cations (denoted by circles labelled 'Cu^{2+}' moving to circles labelled 'Cu') from the electrolyte (represented by blue liquid). Since copper cations are positively charged, their addition makes the cathode positively charged. The anode is 'zinc' in the left beaker, which is shown releasing zinc anions (denoted by circles labelled 'Zn' moving to circles labelled 'Zn^{2+}') into the electrolyte (purple liquid). As zinc metal loses positively charged cations, the anode is negatively charged. The 'salt bridge' is a filter paper soaked in a solution of potassium nitrate (KNO_3), which connects the left and right beakers. Its function is to help maintain the electrical neutrality within the internal circuit. In the image, the nitrate anions (circles labelled 'NO_3^-') move towards the left beaker while the potassium cations (circles labelled 'K^+') move to the right beaker to maintain an electrical neutrality in the two electrolytes. In addition to these key technical elements, the image also involves several technical

DOI: 10.4324/9781003055549-6

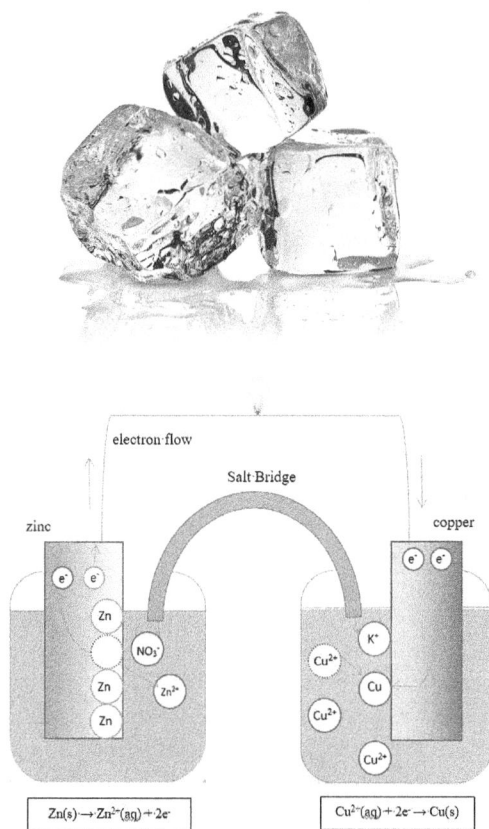

Figure 4.1 Melting ice cubes (top); the electrode reactions in a Daniell cell (bottom). (top) (reproduced with permission from Shutterstock); (bottom) drawing after Chan *et al.* 2018: 391

processes. The two electrodes are associated with two chemical reactions, shown here by labels of chemical equations: the cathode undergoes a reduction reaction '$Cu^{2+}(aq) + 2e^- \rightarrow Cu(s)$' and the anode undergoes an oxidation reaction '$Zn(s) \rightarrow Zn^{2+}(aq) + 2e^-$.' A third technical process is the movement of electrons from the anode to the cathode (denoted by circles labelled 'e$^-$' moving from 'zinc' to 'copper') forming 'electric current.' These processes are overlaid on the elements. In short, the image condenses a significant number of technical meanings from Chemistry.

As the two images from Chemistry textbooks for Years 7 and 11 illustrate, there is considerable difference in the complexity of the knowledge expressed through the years of secondary schooling. This rise in complexity of images is, we shall show, a feature of progression through the years of secondary school Chemistry. Yet, there remains little understanding of how images embody the complexity of Chemistry knowledge. To date, studies have tended to describe images in Chemistry in terms of the kinds of referents they involve. Johnstone (1991), for example, classifies Chemistry knowledge in terms of three levels: 'macro' or what can be seen, touched and smelled; 'submicro' or atoms, molecules, ions, etc.; and 'symbolic' or symbols, formulas, equations, etc. Gilbert (2005) develops this schema into a model of three types of representation in Chemistry: macroscopic, submicroscopic and symbolic. From this perspective, the top image in Figure 4.1 is 'macroscopic' and the bottom image is a hybrid of 'macroscopic' (the apparatus), 'submicroscopic' (atoms, cations, and electrons) and symbolic (chemical equations and formulas). This influential model of chemical representations usefully highlights the breadth of referents that may be included in images, but it does not capture the varying levels of complexity these images present, which is our concern here.

Another approach to images in science education examines the degree of 'specialization' of content expressed (note that this is not 'Specialization' from Legitimation Code Theory). For example, Dimopoulous *et al.* (2003) distinguish three types of images in science: 'realistic' (images that represent reality according to human optical perception), 'conventional' (graphs, maps, flowcharts, molecular structures constructed according to the techno-scientific conventions) and 'hybrids' (images that include elements from both the other two types). They argue that 'conventional' images correspond to strong, 'hybrids' to moderate, and 'realistic' to weak levels of 'specialization.' That is, 'conventional' images express the most techno-scientific knowledge, whereas 'realistic' images convey 'everyday' knowledge. The three categories of images offer a broad sense of differences in images' degree of what they term 'specialization,' which implies differences in the complexity of knowledge. However, it does not systematically capture complexity. For example, an image of a chemical apparatus may be 'realistic' but also express relatively complex technical knowledge. Thus, a model is required for describing different degrees in the complexity of meanings expressed by images.

A fruitful avenue for exploring complexity is through the Semantics dimension of Legitimation Code Theory (LCT). Semantics explores knowledge practices in terms of their context-dependence and complexity (Maton 2011, 2013, 2014, 2020). It has proven useful in analyzing a diverse range of practices, including academic writing (Brooke 2017, Clarence 2017, Kirk 2017), musical performance (Richardson 2020, Walton 2020) and dance (Lambrinos 2020). This chapter will focus on the concept of *semantic density*, which examines the complexity of meanings, and will extend this growing body of work to embrace images. To 'see' the complexity of knowledge expressed by the images used for building Chemistry knowledge, we will establish a model that makes explicit different levels of complexity in images based on data from Chemistry textbooks designed for the secondary school curriculum in NSW, Australia. This model is what is termed in LCT a *translation device* (Maton and Chen 2016) for relating different strengths of semantic density to images from the textbooks. This device will then be enacted to explore how complexity changes through secondary education and to begin to reveal the roles that images play in organizing Chemistry knowledge. Our analysis will suggest that images play a variety of roles, with some connecting knowledge to 'everyday' phenomena and others more concerned with building connections among theoretical ideas. It also suggests that across the years of secondary schooling, at least in textbooks for NSW, images embody a growing range of semantic density. That is to say that, while images in each year maintain a connection to the everyday world, they reach up towards increasingly complex and technical meanings from the field of Chemistry.

Seeing complexity in images: semantic density

Semantic density (SD) refers to the degree of complexity of meanings or practices (Maton 2014). Semantic density can be stronger or weaker along a continuum of strengths, where the stronger the semantic density (SD+), the more complex the meanings and the weaker the semantic density (SD−), the simpler the meanings. Put another way, the more relations with other meanings enjoyed by a practice, the stronger its semantic density (Maton and Doran 2017). These meanings may be epistemological, such as formal definitions and empirical referents, or axiological, such as affective, aesthetic, ethical, moral or political meanings (Maton 2014). In this chapter, we focus on epistemological meanings and so discuss *epistemic–semantic density* or 'ESD.' For example, 'salt' in everyday usage refers to small white crystals often used to add flavour to food – a relatively small number of relations among epistemological meanings, such as its flavour, shape and uses. In contrast, as a technical word in the field of Chemistry, 'salt' refers to a compound produced by the reaction of an acid with a base and involves relations with numerous chemical concepts, such as cations, anions and ionic bonds, which themselves relate to a large number of other meanings. Thus, in Chemistry, the term 'salt' is situated within a relatively complex constellation of epistemological meanings that imbues the term with relatively strong epistemic–semantic density.[1]

As we discussed in the introduction, to analyze the role of the complexity of knowledge expressed by images in building Chemistry knowledge, a model is needed to reveal different degrees of complexity. We begin by outlining such a model as a *translation device* (Maton and Chen 2016) or series of categories for identifying different strengths of epistemic–semantic density in images used in NSW secondary school Chemistry textbooks. We then enact this translation device to explore the complexity of images in Chemistry textbooks through secondary school.

A model of complexity in images used for teaching chemistry

Chemistry uses a range of images, even when representing the 'same' phenomenon. Figure 4.2, for example, presents two images that represent water, from textbooks aimed at Year 7 (left) and Year 11 (right) of secondary school.

The left image comprises water flowing out from the tap into the cup, a depiction of a relatively commonplace activity that shows the physical state and the colour of water. The right image represents water in terms of the intermolecular force among its molecules, using the symbols 'O' and 'H' to indicate atoms within the molecules, solid lines between 'O' and 'H' to represent covalent bonds and a dashed line to represent hydrogen bonding between the water molecules. As our descriptions suggest, the images express different levels of complexity: the left image embodies weaker epistemic–semantic density than the right image, which involves a more complex constellation of meanings and thus stronger epistemic–semantic density.

As shown in Table 4.1, the variation in epistemic–semantic density we touch on here is the first distinction in the model between more complex (stronger ESD) and simpler (weaker ESD) images. It is marked by whether images depend for their meanings on the complex constellation of meanings associated with Chemistry, termed *technical* images, or do not, which we term here *everyday* images.[2] The right image in Figure 4.2 is a *technical* image that makes explicit multiple meanings within the domain of Chemistry, for example, atoms, partial charges and hydrogen bonding. The left image is an

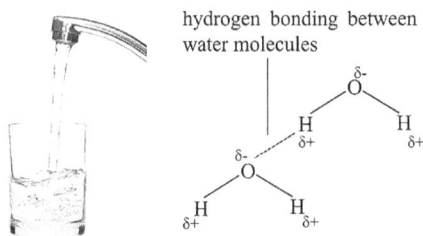

Figure 4.2 Fresh water in a cup (left); hydrogen bonding between water molecules (right).

((left) reproduced with permission from Shutterstock)

Table 4.1 A translation device for the epistemic–semantic density of images in secondary school chemistry textbooks

ESD	Types	Subtypes
++	*technical*	*conglomerate*
		compact
	everyday	*events*
−−		*entities*

everyday image that shows water without indicating a more complex constellation of epistemological meanings. (Although a trained chemist could read into this image a large number of Chemistry-specific meanings, the image itself does not rely on the complex constellation of Chemistry to convey meaning).

Secondary school Chemistry textbooks use both *everyday* and *technical* images throughout year levels to build knowledge. *Everyday* images often present things and phenomena in the physical world that can be connected to specialist Chemistry knowledge, while *technical* images often build theoretical understandings of these things and phenomena.

Subcategories of technical images

This is but a first level of delicacy. Both *everyday* and *technical* images exhibit a range of variations in complexity – as shown in Table 4.1. Focusing first on *technical* images, Figure 4.3 comprises examples from a Year 7 textbook (left), illustrating the compressibility of air, and a Year 11 textbook (right), presenting the structure of water molecules. The left image shows that the process of pushing a piston (the grey stick) compresses air in cylinders (in the right side of the piston), which in turn accelerates the motion of molecules (balls with shades). The image as a whole is categorized as *technical* as it relies on the constellation of Chemistry to make sense. It expresses the knowledge that pressure influences the thermal motion of molecules. However, its various components are not themselves 'technical': the balls, arrows, walls, etc., do not express knowledge from the constellation of Chemistry individually. It is only when combined into the image as a whole that the 'technical' meaning is expressed. In contrast, the right image in Figure 4.3 is categorized as *technical* whose components themselves are also 'technical' in nature. This image is a structural formula that presents the molecular make-up of water: 'O' and 'H' represent oxygen and hydrogen atoms, respectively, and solid lines denote single covalent bonds. The image as a whole is categorized as *technical* as it expresses meanings from the constellation of Chemistry, for example, geometry of water molecules, and the components are 'technical' as they resonate out to numerous meanings in Chemistry, such as lone electrons and shared electrons. Thus, while both are categorized as *technicals*, the Year 11 image embodies stronger

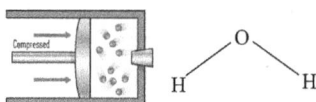

Figure 4.3 Compressibility of air (left); a structural formula of water molecules (right).

((left) reproduced with permission from Oxford University Press)

epistemic–semantic density than the Year 7 image: both the (right) image as a whole and its components are integrated within the complex constellation of meanings that constitutes Chemistry.

This offers a distinction within *technical* images: whether an image involves components that depend for their meanings on complex constellations of epistemological meanings. Those that do are termed *technical-conglomerates*; those where only the image as whole depends on a complex constellation are termed *technical-compacts* (see Table 4.1). In Figure 4.3, the left image is categorized as *compact* – as a whole, it depends on meanings from a wider constellation to express that pressure influences the thermal motion of molecules, but its components do not themselves each express chemical meanings. The right image is categorized as *conglomerate*: both the image as a whole and its components ('O' and 'H,' for example) express chemical meanings.

Though we have limited the device presented in Table 4.1 to two levels of delicacy, for simplicity, further distinctions are possible. For example, Figures 4.4 and 4.5 are both *conglomerates* but represent differences in the complexity of the knowledge they express. Both the images as wholes and their components express chemical meanings. However, Figure 4.5 involves considerably more components that express Chemistry meanings than Figure 4.4. The structural formula of methane molecules (Figure 4.4) involves only two technical components: the chemical symbols representing atoms and the lines representing covalent bonds. In contrast, in Figure 4.5, the two pathways, 'green' and 'brown' (the original is in colour), are themselves technical as they show a series of organic reactions to the target product, ibuprofen molecules. The components within each pathway also express Chemistry meanings. For example, the structural formula of ibuprofen presents its molecular structure. In addition to atoms and covalent bonds, structural formulas in Figure 4.5 involve a group of special technical components: functional groups. For example, the structural formula of ibuprofen molecules includes a carboxylic group (-COOH), which determines that ibuprofen molecules are a type of carboxylic acid. Then within these functional groups are the atoms and the covalent bonds themselves (C, H, – etc.). Thus, further distinctions can be made within *conglomerates* if required. We shall draw on this in our analysis but do not require a third level here for other subcategories.

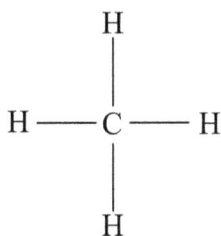

$$H - C - H$$

Figure 4.4 A structural formula of methane molecules.

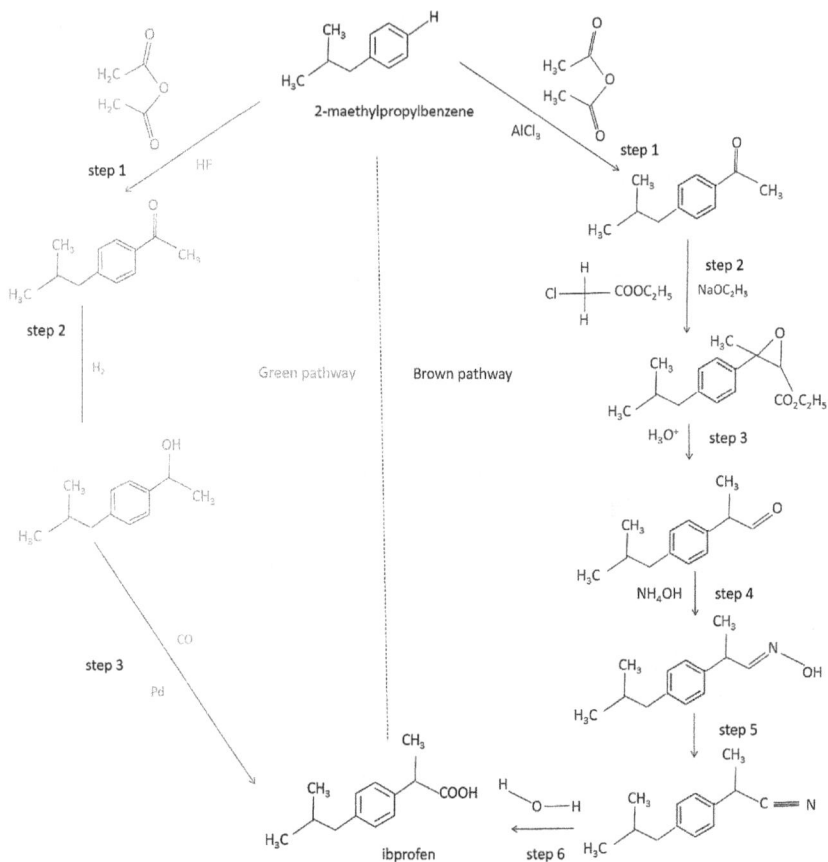

Figure 4.5 Two pathways for the production of ibuprofen.
(drawing after Chan *et al.* 2019: 364)

Subcategories of everyday images

As summarized in Table 4.1, we can distinguish between different kinds of *everyday* images used in Chemistry textbooks. This often involves using *everyday* images for presenting things or phenomena as they appear to the naked eye. Such images often offer an orientation to the phenomena explained by Chemistry as they appear in commonplace settings. A second level of delicacy is illustrated by Figure 4.6. The left image shows the inside of a box of matches, while the right image displays a burning match. Both are *everyday* images but vary in complexity. The left image shows entities: the box and matches. The right image shows both an entity and an event: the burning of a match. The left image thus displays weaker epistemic–semantic density than the right image, which includes a process. We can thereby distinguish between: *everyday-entity* images (ESD−−) that present entities only and *everyday-event* (ESD−) images that also express events or actions.

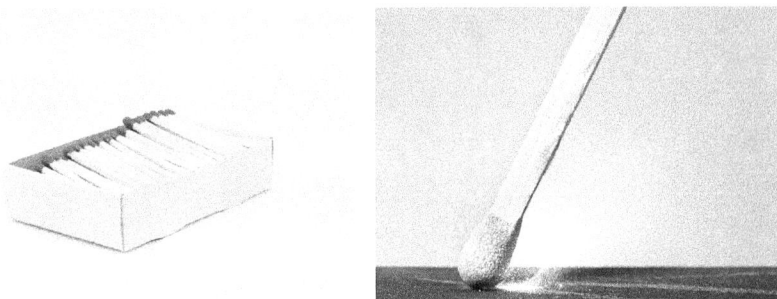

Figure 4.6 A pile of matches (left); a burning match (right).

((left) reproduced with permission from Shutterstock; (right) reproduced with permission from Shutterstock)

Distinctions among images and knowledge-building

The distinction between different kinds of *everyday* and *technical* images allows us to 'see' the potential for knowledge-building offered by the sequencing of images in Chemistry textbooks. Using the model outlined above, we can see how the 'same' Chemistry phenomenon can be imaged with varying degrees of complexity. For example, all three images in Figure 4.7 represent combustion of carbon; the left and middle images are from Year 10, and the right is from Year 11. The top image presents combustion of carbon through an *everyday-event* image (ESD–): a pile of wood is burning at a campsite to provide light and warmth for people. The left image presents combustion of carbon through a *technical-compact* image (ESD+) that describes energy change during the reaction. The vertical axis shows energy levels of chemical species and the horizontal axis indicates reaction time. The line graph shows that, as the reaction goes on, energy decreases from a higher level to a lower level. The image as a whole relates to the chemical concept of exothermic reaction, which explains why the burning of wood in the left image can release heat and warm people. With stronger epistemic–semantic density, the middle image provides a theoretical explanation for the physical phenomenon presented in the left image. The knowledge of energy change during combustion of carbon is then further elaborated in Year 11 through a *technical-conglomerate* image (ESD++), shown in the right image in Figure 4.7. This is an energy level diagram for the formation of carbon dioxide from carbon and oxygen via carbon monoxide. The diagram shows two pathways to the formation of carbon dioxide, in which the overall enthalpy change involved is the same. The image as a whole is 'technical' as it expresses meanings from the constellation of Chemistry: combustion of carbon to form carbon dioxide involves two stages and releases heat. Within the diagram, the components also express 'technical' meanings. The arrows that lead '$C(s) + O_2(g)$' to '$CO(g) + \frac{1}{2} O_2(g)$' and '$C(s) + O_2(g)$' to '$CO_2(g)$' represent incomplete and complete combustion of carbon, respectively, and are meanings from the constellation of Chemistry. With even stronger epistemic–semantic density, the

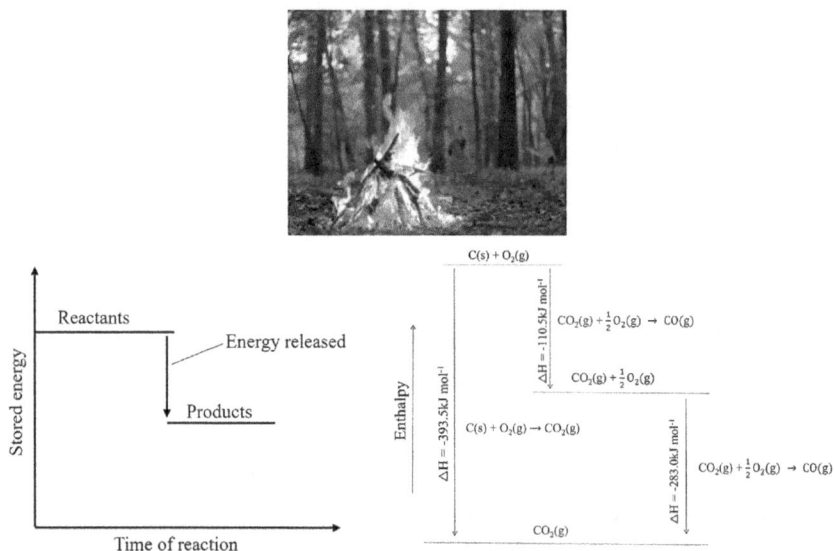

Figure 4.7 A campfire (top); energy change during combustion of carbon (left); an energy level diagram for the formation of carbon dioxide from carbon and oxygen carbon monoxide (right).

((top) reproduced with permission from Shutterstock; (right) drawing after Chan *et al.* 2018: 486)

right image expresses more complex Chemistry knowledge of the combustion of carbon.

Though the three images in Figure 4.7 represent the 'same' phenomenon, they present differing degrees of complexity of meaning. The *everyday* image presents a 'common-sense' or experiential version of the phenomenon and the *technical* images offer increasingly complex theoretical understandings. When sequenced together, they offer the possibility for connecting simpler meanings to a greater range of meanings from the constellation of Chemistry, increasing the complexity of the knowledge being expressed. To explore how images are used in Chemistry textbooks, we shall enact the translation device (Table 4.1) to analyze images from Chemistry textbooks designed for the secondary school curriculum in NSW, Australia.

Changing complexity of images in chemistry textbooks

Chemistry textbooks tend to use images with a range of different levels of complexity both within and across year levels. To explore this variation, we will use the translation device for epistemic–semantic density to examine Chemistry textbooks designed for secondary schooling in NSW, Australia. Here we analyze six textbooks: Oxford Insight Science Year 7 (Zhang *et al.* 2013), Oxford Insight Science Year 8 (Zhang *et al.* 2014a), Oxford Insight Science Year 9 (Zhang *et al.* 2014b), Oxford Insight Science Year 10 (Zhang *et al.* 2015), Pearson Chemistry Year 11 (Chan *et al.* 2018) and Pearson Chemistry Year 12 (Chan *et al.* 2019). It should be noted that our analysis of images excludes the accompanying text in the textbooks; our focus here is on what images themselves express in terms of complexity of knowledge, rather than their multimodal interactions with verbal text.

The curriculum of secondary school Chemistry in NSW includes six years that are categorized into three stages: Stage 4 (Years 7 and 8), Stage 5 (Years 9 and 10) and Stage 6 (Years 11 and 12). Our analysis focuses on one topic that appears in all three stages: chemical reactions. We shall show that while epistemic–semantic density increases across the stages, weaker epistemic–semantic density images are present throughout. Thus, rather than a progression from simpler to more complex images, the textbooks retain simpler images and add more complex images. That is, simpler images are not confined to an early stage of secondary schooling and left behind once technical Chemistry knowledge has been established. Rather, a connection to the 'everyday' and experiential is kept from Year 7 to Year 12. In sum, they represent a growing *range* of epistemic–semantic density. We suggest that this offers textbooks the possibility of modelling moves between simpler 'everyday' or common-sense knowledge and more complex theoretical Chemistry knowledge throughout secondary school.

Stage 4: From everyday-entities to technical-compacts

Figure 4.8 shows the simplest and the most complex images found in the textbooks for Stage 4 (Year 7 and 8) in relation to chemical reactions. They are from consecutive pages of the same textbook. The left image shows a rusted car, illustrating a common issue for objects made of iron: rust. This is an *everyday-entity* image that expresses relatively weak epistemic–semantic density (ESD– –). In contrast, the right image shows a diagram of six test tubes containing different environments: air, water, oil, boiled water, salt solution and dry salt. In these test tubes, iron nails are placed to observe their respective speed of rusting. This diagram thus also illustrates 'rusting' but

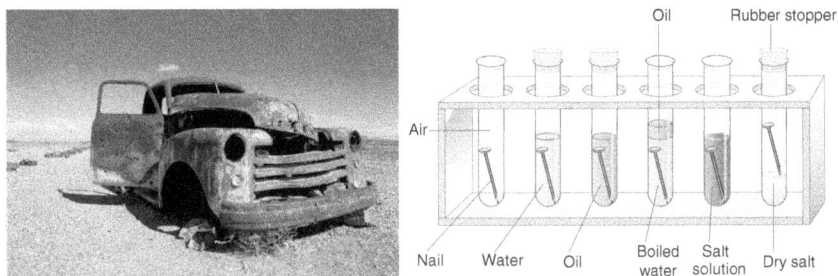

Figure 4.8 A rusted car (left); a diagram of a rusty nail experimental set-up (right).

((left) reproduced with permission from Shutterstock; (right) reproduced with permission from Oxford University Press)

with a *technical-compact* image that expresses stronger epistemic–semantic density (ESD+). The six test tubes include different environments with distinct concentrations of air, which indicates that the key variable influencing rusting of iron is oxygen. With this range of epistemic–semantic density (from ESD–– to ESD+), the images in Stage 4 offer imagic means to everyday phenomena and build relatively theoretical understanding underpinning them (though not as technical as is possible: ESD+ rather than ESD++).

Stage 5: From everyday-entities to technical-conglomerates

Figure 4.9 shows the simplest and the most complex images in relation to chemical reactions found in textbooks for Stage 5 (Years 9 and 10). Similarly to Stage 4, there is an *everyday-entity* image that expresses relatively simple meanings (ESD––): on the left image in Figure 4.9, showing two segments of orange. This is illustrating a food that has weak acidity and reactivity and thus is edible. In contrast to Stage 4, however, the kind of *technical* images included in textbooks are of greater complexity: *technical-conglomerate* images (ESD++). The right image of Figure 4.9 illustrates the formation of sodium chloride through an 'equation diagram.' The overall diagram shows that a sodium chloride (Na^+ and Cl^-) is formed by a sodium atom (Na) donating an electron to a chlorine atom (Cl) (shown by the dashed arrow combined with an 'e^-'). In addition, the components – the individual diagrams showing atomic structure of the atoms and ions – are technical. One aspect of the key technical information expressed by the atomic structure diagram is that a sodium atom has only one electron in its outer shell and a chlorine atom has seven. A sodium atom and a chlorine atom thus tend to lose and gain one electron, respectively, to achieve their stable status. This *technical-conglomerate* means that epistemic–semantic density of images in Stage 5 reaches higher than in Stage 4: it reaches further into complexity. However, as shown by the left image, images expressing simpler meanings are retained: everyday-entity images (ESD––). Thus, images in Stage 5 span a greater range of epistemic–semantic density (from ESD–– to ESD++) than those found in Stage 4 (from ESD–– to ESD+).

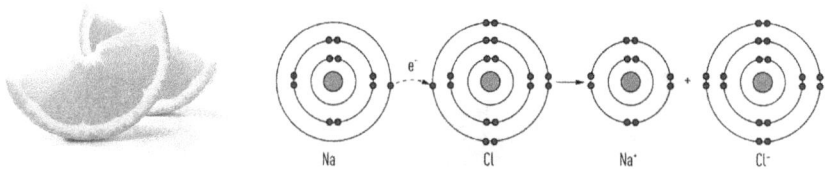

Figure 4.9 Two segments of orange (left); the formation of sodium chloride.

((left) reproduced with permission from Shutterstock; reproduced with permission from Oxford University Press)

Stage 6: From everyday-entities to technical-superconglomerates

Stage 6 takes a further step into complexity. The textbooks continue to include *everyday-entity* images but now include far more *technical-conglomerate* images and introduce particularly strong *technical-conglomerates* or, for want of a better term for the moment, *superconglomerates*. The range of epistemic–semantic density of images is illustrated by Figure 4.10. The left image is an *everyday-entity* (ESD--) that shows batteries and is used in the textbook to illustrate the application of electrochemical reactions in an 'everyday' setting. The right image is a *technical-superconglomerate*. It shows the formation of a secondary amide through a condensation reaction between ethanoic acid and methylamine. The diagram as a whole is 'technical' as it shows a Chemistry reaction integrated within the domain of Chemistry. Each side of the arrow is also 'technical.' The left includes diagrams known as structural formulas for both ethanoic acid and methylamine, while the right includes structural formulas for water and a secondary amide. These components each in turn include multiple other 'technical' components. For example, in the structural formula of ethanoic acid, the group '–COOH' represents the functional group carboxyl, while '–CH$_3$' represents the methyl group, with each of these further including technical components: the symbols 'C,' 'O,' and 'H' represents atoms, while the lines '—' and ' ' denote single and double covalent bonds. This image thus expresses multiple levels of technical meaning – more than that shown by the most complex images of Stages 4 or 5. This represents significantly stronger epistemic–semantic density. Such images occur regularly in Stage 6 and thus push the complexity of the knowledge expressed through images to much greater heights.

A greater range of complexity

Textbooks for all three Stages of NSW secondary schooling employ *everyday-entity* images (ESD--) to show 'everyday' phenomena. However, as illustrated in Figure 4.11, there is an expansion in the degree of complexity of other images as the years progress, increasing the semantic range of images. As Figure 4.11 suggests, the knowledge expressed by the images in the textbooks studied maintains connections with the common-sense 'everyday' or

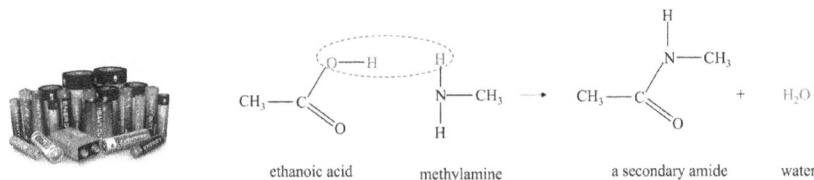

Figure 4.10 Batteries (left); a diagram showing the formation of a secondary amide through a condensation reaction between ethanoic acid and methylamine (right).

((left) reproduced with permission from 123RF.com)

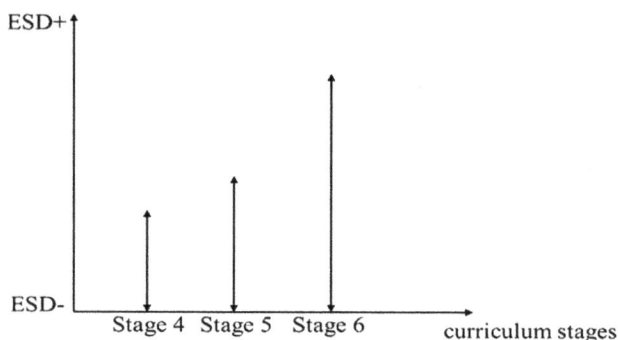

Figure 4.11 A widening range of images' epistemic–semantic density across stages.

phenomenal world but reaches towards increasing levels of complexity in Chemistry knowledge. As we have emphasized, this is not simply an ascent into greater complexity but rather into a greater *range of complexity*.

We conjecture that this growing range of epistemic–semantic density may play a significant role in building Chemistry knowledge through secondary schooling. In terms of Johnstone's (1991) Chemistry triplet (macro, micro and symbolic), the simplest images in each stage tend to express macroscopic knowledge (what can be sensed) and present things or phenomena as they appear in everyday settings. As the epistemic–semantic density range grows, the most complex images tend to express microscopic knowledge (diagrams of atoms, molecules and structures) in Stage 5 and symbolic knowledge (symbolic graphs and diagrams) in Stage 6. In the field of Chemistry education, it has been widely argued that to be successful students need to move among different types of Chemistry knowledge (Gabel 1993, Johnstone 1993, Chittleborough *et al.* 2005). The widening range of epistemic–semantic density of images in textbooks analyzed here suggests that this move, particularly when shifting from macroscopic knowledge to symbolic knowledge, involves successful mastery of the increasing range of complexity of the knowledge expressed by images.

Integrating complexities through composite images

This range of complexity is not simply found in comparing images – it may also be expressed *within* images. To explore this, we introduce another form of images: *composites*. These are images that bring together different degrees of complexity of knowledge. For example, Figure 4.12 is an image that brings together two levels of complexity representing water and ice. The photograph on the left shows ice floating on water, a common-sense phenomenon we could see in 'everyday' life. The two diagrams, in contrast, provide a chemical explanation at a microscopic level for the phenomenon: ice floats on the water because the water molecules form a crystal lattice in which the molecules are spaced more widely apart than in liquid water. This arrangement of water molecules means ice is less dense than liquid water.

These *everyday* and *technical* images embody two levels of complexity and are incorporated into one *composite*.

A *composite* image can be distinguished by considering two attributes. First, the image presents strong boundaries between its composite parts.[3] These boundaries tend to be shown by elements which create dividing lines that disconnect constituent parts of the image or by blank spaces that form 'gutters' or gaps between images. Both signify that there are constituent parts being brought together. In Figure 4.12, the circles which centre around the molecule models separate the photo and the two diagrams. Viewed individually, each of these constituent images makes sense on their own. Second, the constituent images of a *composite* embody different levels of complexity. As mentioned above, Figure 4.12 comprises images exhibiting two levels of complexity: the photograph of a cup of water and ice embodies weaker epistemic–semantic density (an *everyday-entity*), and the diagrams of water molecules embody relatively strong epistemic–semantic density (*technical-conglomerate*). By way of contrast, Figure 4.13 neither distinguishes constituents nor involves different levels of complexity. Though it may look as if it contains multiple components, they do not have strong boundaries around each other nor embody distinct strengths of epistemic–semantic density.

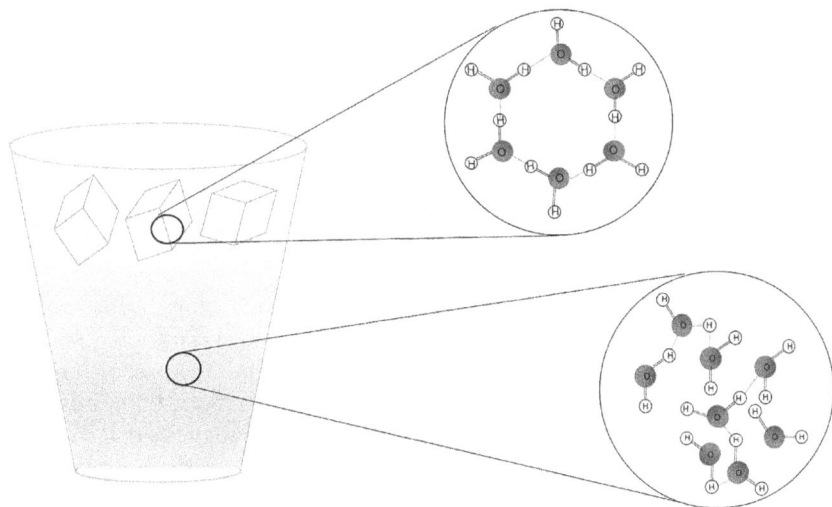

Figure 4.12 Ice is less dense than liquid water.
(drawing after Chan *et al.* 2018: 183).

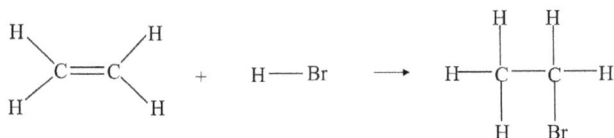

Figure 4.13 The addition reaction of ethene with bromine (Chan *et al.* 2019: 313).

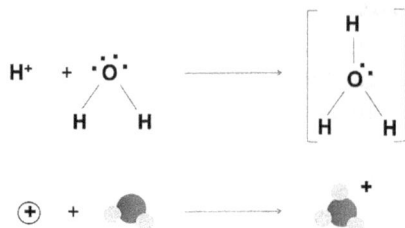

Figure 4.14 The formation of the hydronium ion.
(drawing after Chan *et al.* 2019: 148)

The key point for our analysis here is that *composites* offer a way of bringing together a range of complexities in one eyeful. Significantly for our conjecture of a growing range of epistemic–semantic density through secondary schooling, *composite* images in textbooks occur mainly in Stage 6 (though remain relatively infrequent compared to non-composite images).

In addition to *composites* that combine *everyday* and *technical* images, composites can also bring together *technical* images of different levels of complexity. For example, Figure 4.14 (from Year 12) brings together: a *technical-compact* representation of a model diagram on the bottom showing a water molecule (the diagrammatic model consisting of one red ball and two smaller white balls) adopting a proton (the white ball labelled '+') to become a hydronium ion (the diagrammatic model on the right of the arrow) and a *technical-conglomerate* representation on the top that represents the reaction through a diagrammatic equation. Arranging these images together relates Chemistry meanings that are themselves complex to even more complex meanings.

The *composites* in Figures 4.12 and 4.14 offer a way of 'bridging' between meanings with different degrees of complexity. They embody a range of epistemic–semantic density. This may help compound meanings from simpler into complex forms. In terms of Johnstone's (1991) chemical triplet, *composite* images present a transition between macroscopic and microscopic knowledges in Figure 4.12 and between microscopic and symbolic knowledges in Figure 4.14. The change in the complexity of the images suggests that students are expected to be able to understand the shift between the different levels of epistemic–semantic density of the images embodied by the *composites*.

Conclusion

Along with language and chemical formalisms, images are one of the key means of expressing Chemistry knowledge in textbooks and beyond. Images may exhibit different degrees of complexity in different learning stages. However, to date, there has been little exploration of how images embody the complexity of Chemistry knowledge. This chapter has explored the complexity of images using the concept of epistemic–semantic density from LCT and developed a model for complexity of images as used in secondary school Chemistry text-books in NSW, Australia. Enacting the device, we suggested that to build Chemistry knowledge, images develop from the *everyday* category with rela-tively weak epistemic–semantic density to the *technical* category with increas-ingly stronger epistemic–semantic density. The former tends to play the role of presenting common-sense physical and experiential phenomena, thereby offer-ing students a common-sense 'way into' understanding, whereas the latter usu-ally offers theoretical understandings of the phenomena.

Our analysis of the range of epistemic–semantic density of images across curriculum stages suggests that knowledge expressed by the images in each stage maintains connections with the 'everyday' world but shows an increas-ing complexity as the curriculum progresses. This widening range of epis-temic–semantic density indicates that in each stage students are expected to engage with both *everyday* and *technical* images. In Stage 6, the textbooks additionally involve *composite* images that bring together images embodying different levels of complexity, either between *everyday* images and *technical* images or between *technical* images with different degrees of complexity. This suggests that students are expected to be able to move between the different levels of complexity to connect meanings together.

The translation device for epistemic–semantic density presented in this chapter was developed through analysis of textbooks. However, it is likely it will be of use to images in Chemistry more broadly, such as images used in classroom teaching or assessments. Although in LCT terms, they are *specific translation devices* for the problem-situation of a specific study (see Maton and Howard 2018), they offer a pathway to developing a *generic translation device* that works for all images in Chemistry. Our chapter is also limited to offering a way of seeing the complexity of knowledge expressed by images in Chemistry education. How this can be enacted to support teaching with images is, as yet, unexplored. Nonetheless, the model offered here represents, we believe, a valuable first step towards seeing the complexity of images in Chemistry.

Notes

1 On *constellations* in LCT, see Maton (2014) and Maton and Doran (2021). Here we are only concerned the complexity of network of relations among meanings within which a meaning is situated.

2 We have as far as possibly echoed the names of categories in the *generic transla-tion device* for English discourse as a whole set out by Maton and Doran (2017) in order to facilitate inter-modal comparison and multimodal analysis in future. However, English discourse and images are different objects of study whose dif-ferent attributes often required different labels. We should emphasize that the current paper offers a *specific translation device* for images in secondary school chemistry textbooks, rather than a *generic translation device* for all images in Chemistry, let alone all images.

3 Kress and van Leeuwen (2020) call these boundaries 'framing,' a term which is easily confused with a different meaning of 'framing' by Bernstein (1973). We shall not use the term here.

References

Bernstein, B. (1973) *Class, codes and control volume 1: Theoretical studies towards a sociology of language.* Routledge, Kegan and Paul.

Brooke, M. (2017) 'Using "semantic waves" to guide students through the research process: From adopting a stance to sound cohesive academic writing,' *Asian Journal of the Scholarship of Teaching and Learning,* 7(1), 37–66.

Chan, D., Commons, C., Hecker, R., Hillier, K., Hogendoorn, B., Lennard, L., Moylan, M., O'Shea, P., Porter, M., Sanders, P., Sturgiss, J. and Waldron, P. (2018) *Pearson chemistry 11 New South Wales student book.* Pearson Australia.

Chan, D., Commons, C., Commons, P., Finlayson, E., Hillier, K., Hogendoorn, B., Johns, R., Lennard, L., Moylan, M., O'Shea, P., Porter, M., Sanders, P., Sturgiss, J. and Waldron, P. (2019) *Pearson Chemistry 12 New South Wales student book.* Pearson Australia.

Chittleborough, G., Treagust, D. and Mocerino, M. (2005) 'Non-major chemistry students' learning strategies: Explaining their choice and examining the implications for teaching and learning,' *Science Education International,* 16(1): 5–21.

Clarence, S. (2017) 'A relational approach to building knowledge through academic writing: Facilitating and reflecting on peer writing tutorials,' in S. Clarence and L. Dison (eds) *Writing Centres in Higher Education: Working in and across the Disciplines.* SUN Press.

Dimopoulos, K., Koulaidis, V. and Sklaveniti, S. (2003) 'Towards an analysis of visual images in school science textbooks and press articles about science and technology,' *Research in Science Education,* 33(2): 189–216.

Gabel, D. L. (1993) 'Use of the particulate nature of matter in developing conceptual understanding,' *Journal of Chemical Education,* 70(3): 193–194.

Gilbert, J. K. (2005) 'Visualization: A meta-cognitive skill in science and science education,' in J.K. Gilbert (ed.) *Visualization in Science Education.* Springer.

Johnstone, A. H. (1991) 'Why is science difficult to learn? Things are seldom what they seem,' *Journal of Computer Assisted Learning,* 7(2): 75–83.

Johnstone, A. H. (1993) 'The development of chemistry teaching: A changing response to changing demand,' *Journal of Chemical Education,* 70(9): 701–705.

Kapıcı, H. Ö. and Savaşcı-Açıkalın, F. (2015) 'Examination of visuals about the particulate nature of matter in Turkish middle school science textbooks,' *Chemistry Education Research and Practice,* 16(3): 518–536.

Kirk, S. (2017) 'Waves of reflection: Seeing knowledge(s) in academic writing,' in *EAP in a Rapidly Changing Landscape: Issues, Challenges and Solutions. Proceedings of the 2015 Baleap Conference.* Reading: Garnet Education, 109–117.

Kress, G. and van Leeuwen, T. (2020) *Reading Images: The Grammar of Visual Design.* 4th ed. Routledge.

Lambrinos, E. M. (2020) 'Building ballet: Developing dance and dancers in ballet,' Unpublished PhD thesis, University of Sydney, Australia.

Maton, K. (2011) 'Theories and things: The semantics of disciplinarity,' in F. Christie and K. Maton (eds) *Disciplinarity: Functional Linguistic and Sociological Perspectives.* Continuum.

Maton, K. (2013) 'Making semantic waves: A key to cumulative knowledge-building,' *Linguistics and Education,* 24(1): 8–22.

Maton, K. (2014) *Knowledge and Knowers: Towards a Realist Sociology of Education.* Routledge.

Maton, K. (2020) 'Semantic waves: Context, complexity and academic discourse,' in J.R. Martin, K. Maton and Y.J. Doran (eds) *Accessing Academic Discourse: Systemic Functional Linguistics and Legitimation Code Theory*, Routledge.

Maton, K. and Chen, R. T-H. (2016) 'LCT in qualitative research: Creating a translation device for studying constructivist pedagogy,' in K. Maton, S. Hood and S. Shay (eds) *Knowledge-Building: Educational Studies in Legitimation Code Theory*, Routledge.

Maton, K. & Doran, Y. J. (2017) 'Semantic density: A translation device for revealing complexity of knowledge practices in discourse, part 1 – wording,' *Onomázein*, Special Issue, March: 46–76.

Maton, K. & Doran, Y. J. (2021) 'Constellating science: How relations among ideas help build knowledge,' in K. Maton, J.R. Martin and Y.J. Doran (eds) *Teaching Science: Knowledge, Language, Pedagogy*, Routledge.

Maton, K. and Howard, S. K. (2018) 'Taking autonomy tours: A key to integrative knowledge-building,' *LCT Centre Occasional Paper* 1: 1–35.

Pintó, R. and Ametller, J. (2002). 'Students' difficulties in reading images: Comparing results from four national research groups.' *International Journal of Science Education*, 24(3): 333–341.

Richardson, S. (2020) 'Teaching Jazz: A study of beliefs and pedagogy using Legitimation Code Theory,' Unpublished PhD thesis, University of Sydney, Australia.

Walton, J. (2020) 'Making the Grade: Theorising musical performance assessment,' Unpublished PhD thesis, Griffith University, Australia.

Zhang, J., Alford, D., McGowan, D. and Tilley, C. (2013) *Oxford Insight Science 7 Student Book*, Oxford University Press.

Zhang, J., Alford, D., Hopley, S. and Tilley, C. (2014a). *Oxford Insight Science 8 Student Book*, Oxford University Press.

Zhang, J., Alford, D., Morante, R. and Tilley, C. (2014b). *Oxford Insight Science 9 Student Book*, Oxford University Press.

Zhang, J., Filan, S.D., Hopley, S., Morante, R. and Tilley, C. (2015). *Oxford Insight Science 10 Student Book*, Oxford University Press.

5 Using variation in classroom discourse

Making Chemistry more accessible

Bruno Ferreira dos Santos, Ademir de Jesus Silva Júnior, and Eduardo Fleury Mortimer

Introduction

Traditional practices for teaching Science, particularly the teaching of Chemistry, are charged with alienating students from Science by representing Chemistry knowledge in an abstract way, far removed from and without relevance to students' everyday lives (Stuckey *et al.* 2013). In addition, these traditional ways of teaching Chemistry generally ask the learners to be passive in the classroom, as listeners, characterizing what Paulo Freire coined as a 'banking' model of education (Freire 2018). Chemistry school courses are also viewed as fact-oriented and memorization-oriented, unattractive to many students who are more inclined to creative activities, in contrast to what the curricula may offer them. Furthermore, the prominence of mathematically based problems to test knowledge of Chemistry content does not indicate a real understanding of chemical concepts by the students but only the use of memorized algorithms (Nakhleh 1993).

Chemistry is considered a difficult subject to teach and one reason for this is that Chemistry knowledge can be described at several levels or domains, only one of which can be observed (Nelson 2002). It was Alex Johnstone (1982, 1993, 2000, 2003) who introduced this concern about the nature of Chemistry knowledge in the field of chemical education. Johnstone (1982) treated these levels as formal aspects of Chemistry teaching and designated them as follows: the *macro* (dealing with experiences and observations of concrete substances and phenomena), the *symbolic* or *representational* (dealing with symbols, equations, and calculations) and the *molecular* or *submicro* (dealing with molecules, atoms, structure and bonding, see Figure 5.1). This multi-level structure is also known as the 'triplet relationship' or Johnstone's chemical triangle (Chittelborough 2014; Sjöström and Talanquer 2014; Talanquer 2011; Gilbert and Treagust 2009).

When students are introduced to Chemistry and other scientific subjects in school, they are introduced to a highly specialized discourse, a shift that moves them from common sense to an 'uncommon sense understanding of the world' (Martin 2013: 23). Mastering knowledge of Chemistry in an appropriate way requires that students connect and reason across the three levels of the triplet relationship, which may be difficult and confusing for

DOI: 10.4324/9781003055549-7

PHYSICAL SCIENCES

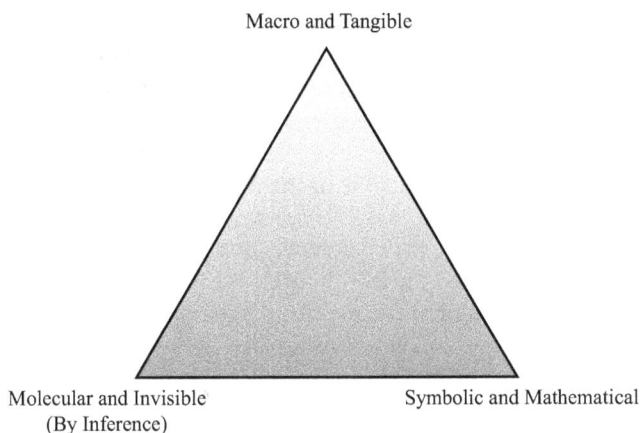

Macro and Tangible

Molecular and Invisible
(By Inference)   Symbolic and Mathematical

Figure 5.1 The triplet relationship.

(adapted from Johnstone 2010)

them (Liu and Taber 2016; Becker *et al.* 2015; Chittelborough 2014; Stieff *et al.* 2013; Wu 2003). If we agree that learning Chemistry can be viewed as the acquisition of its particular language, the nature of this language is complex and its representations in several semiotic modes are a challenge for students (Danielsson *et al.* 2018). The recognition of all of these issues brought about the study of scientific language acquisition to trace students' learning to the agenda of scientific education.

Since 1990, there is a growing interest in the use of language in Science classrooms among Science education researchers, which results in a number of different approaches to classroom discourse (Fensham 2004). In these studies, researchers try to understand how the different patterns of discursive and non-discursive interactions between teachers and students build the classroom communication processes that represent the scientific knowledge, skills and values of Science curricula. Students' participation in these interactions and their written productions constitute the sources to analyze their appropriation of the language of Science.

Although these recent perspectives are broadening the ways we comprehend the teaching and learning of Science, most classroom-focused studies draw on the interactional domain, and researchers seek structures or patterns of discursive interactions that build social relations through the use of language. This trend indicates that the research on classroom discourse shares a pervasive characteristic of the research in education: *knowledge-blindness* (Maton 2014a). Maton argues that, regardless of the knowledge being the basis of education, a subjectivist *doxa* in education research reduces knowledge to knowing or to power. The knowledge-blindness means that under the influence of this subjectivist *doxa*, knowledge as an object is underresearched and the study of education underdeveloped. For Maton, knowledge is not a homogeneous and neutral object but can assume various forms and have different effects.

According to Maton, taking knowledge seriously requires 'the right conceptual tools for analyzing this object of study' (Maton 2014a: 13). He developed Legitimation Code Theory (LCT) as a multidimensional conceptual toolkit and analytic methodology that 'enables knowledge practices to be seen, their organizing principles to be conceptualized, and their effects to be explored' (Maton 2014a: 2). One of the dimensions of LCT is Semantics. The key concepts of Semantics are *semantic gravity* and *semantic density*, which conceptualize context-dependence and complexity of knowledge practices (Maton 2013, 2014a, 2014b, 2020).

Semantic density (SD) provides us with the means to explore 'how meanings are condensed and interrelated within knowledge practices' (Maton and Doran 2017: 47). According to Maton (2013), the strength of semantic density can vary along a continuum, where meanings are added or 'packed' to symbols or practices through a process called *condensation*: the stronger the semantic density (SD+), the more meanings are condensed within practices; the weaker the semantic density (SD–), the fewer meanings are condensed. The movement to weaken the semantic density is called *rarefaction*, whereby the meanings are 'unpacked' or removed (Maton 2014a). Maton (2013) asserts the condensation of meanings around concepts and empirical referents constitutes *epistemological condensation*, while *axiological condensation* involves emotional, political, ethical and moral stances.

Semantics also provides a way to study *cumulative knowledge-building*, defined by Matruglio *et al.* as movement 'which involves both looking backwards to previous ideas and looking forwards to future contexts in which current knowledge can be applied and extended' (Matruglio *et al.* 2013: 38). A key characteristic of knowledge-building through text-based classroom discourse is a *semantic wave*, defined by Maton as 'recurrent shifts in context dependence and condensation of meaning that *weave* together different forms of knowledge' (Maton 2014b: 182).

Taking into account all of the issues associated with the teaching and learning of Chemistry, it is surprising that very few studies drawing on classroom discourse have explored the structure of the knowledge represented in the discourse. Instead of the structure of the knowledge, most research prospect characteristics of 'knowing' Chemistry. This feature follows the majority trend in Science education research, which focuses on students' ideas or conceptions (Georgiou 2016).

In agreement with Blackie (2014) and Rootman-le Grange and Blackie (2018), who viewed a fruitful framework for thinking about Chemistry education through Semantics, we consider that this theory enables us to examine the complexity of scientific knowledge for the purpose of understanding the way this object can be built through talk and instruction within the discourse of a Chemistry classroom. In this chapter, we focus on one organizing principle in Semantics – *semantic density* – and we explore the variation of semantic density within the discourse of Chemistry classes as an expression of the complexity of the knowledge in the teaching practice. For this, we developed and present what LCT terms 'translation devices' for enacting the concept of semantic density in analysis of Chemistry classroom discourse. The role of translation devices is to relate abstract concepts, such as semantic density, to specific empirical data (Maton and Chen 2016). We use our translation devices in the analysis of the variation of semantic density in the classroom discourse of four high school Chemistry teachers. We discuss the scope and limitations of the translation devices based on the nature of our data and the results of the analysis. Beyond exploring the knowledge practices in the Chemistry classroom discourse, we hope our chapter also contributes to expanding the number of characteristics revealed in social practices when semantic concepts are enacted within specific contexts of substantive studies (Maton 2013).

The chapter is organized as follows: first, we outline our translation devices for enacting semantic density to analyze Chemistry classroom discourse; second, we describe the data, including the sample and educational contexts where we gathered our data; third, we illustrate our analysis with some episodes of Chemistry classroom discourse; and, fourth, we discuss our results, drawing on our theoretical framework.

The translation devices

Semantic concepts are being applied to a growing number of studies that involve different phenomena. To be used on such diverse data and practices, it is necessary to develop translation devices 'for translation between the concepts and their differing realizations within specific objects of study' (Maton 2013: 21).

Regarding semantic density, Maton and Doran (2017) argue that a translation device for educational discourse must allow for analysis of strengthening and weakening of semantic density and capture the differences and movements in such a way that we can trace the variation of semantic density within practices over time. The expected variation of semantic density gives us the semantic density profile and the semantic density range between its highest and lowest strengths (Maton 2014b).

As we are interested in exploring the different forms assumed by *epistemological condensation* in classroom discourse, our translation devices enact *epistemic–semantic density* (ESD). Our first translation device developed to enact ESD in analysis of classroom discourse is shown in Table 5.1. This device (Santos and Mortimer 2019) describes four levels that represent relative strengths of ESD. It relates to the organization of chemical knowledge along three dimensions, as represented by the triplet relationship (Talanquer 2011). Stronger ESD corresponds to the symbolic domain and, in classroom discourse, this level consists of all talk for which the object of discussion is a chemical symbol, diagram, graph or image. Weaker ESD corresponds to talk for which the object is situated in the macroscopic or phenomenological domain; that is, those objects tangible to us. Nevertheless, we divide this

Table 5.1 Translation device for epistemic–semantic density of chemistry knowledge

Semantic density	Level	Form	Description	Example
Stronger ↕ Weaker	ESD++	Symbolic	Chemical symbols, diagrams, graphs, images	Diagram showing change of state
	ESD+	Submicroscopic conceptual	Demands the understanding of corpuscular theory to explain the phenomena	Association between the boiling temperature of a liquid and its molecular properties
	ESD−	Macroscopic conceptual	Relates scientific concepts with macroscopic aspects of the phenomena	Association between evaporation and boiling temperature of a liquid
	ESD−−	Macroscopic phenomenological	Relates concepts used in daily language regarding the phenomena	Association between the evaporation of a liquid with the empiric description of the observation

Adapted from Santos and Mortimer (2019)

level into two different levels. At level one (ESD--), the phenomena are described using only everyday language, whereas at level two (ESD-), the described phenomena are condensed by using concepts that introduce scientific language. We think that the use of scientific language adds complexity to the description of a phenomenon, as the scientific concepts have precise definitions and introduce new classification and composition structures to the description, resulting in a higher level of ESD. Level three (ESD+) corresponds to the domain of submicroscopic entities of Chemistry, demanding the corpuscular theory of atoms, molecules and ions and other particles to describe and interpret the phenomena. When using the atomic-molecular theory, it is necessary to introduce a series of new entities that increases the complexity of the explanation, strengthening ESD. The next level (ESD++), corresponding to the symbolic representation, adds still more complexity to the explanation as now, beyond the models, one should deal also with the representation of the models.

We used this tool to draw profiles of semantic density in the analysis of and contrast between the instruction discourses of two Chemistry school teachers (Santos and Mortimer 2019). The inclusion of level four in this device is under review since we are now considering that the symbolic domain demands a translation device of its own (see Chapter 4, this volume). Because of this, we will not analyze data in this chapter that comprise symbols, images or diagrams, only spoken discourse.

Our first translation device is strongly attached to the features of Chemistry knowledge. However, our incursion of LCT in the study of Chemistry teaching resulted in a need to identify 'markers' in the classroom discourse that promote the shifts in the strength of SD. We first called upon *pedagogical link-making*, a set of ways 'teachers and students make connections between ideas in the ongoing meaning-making interactions of classroom teaching and learning' (Scott *et al.* 2011: 5). In addition, we traced the different shapes of the semantic density profiles to the teachers' discourse which revealed that we also needed to complement the analysis of ESD with a device that allowed us a finer immersion in the language of the classroom discourse.

To construct our second translation device, we drew on Maton and Doran (2017) and their series of categories for *wording* and *wording-group*. By means of these tools, we explored how different strengths of ESD express different degrees of knowledge complexity at the word level. The wording tool is shown in Table 5.2 and focuses on 'content words' such as nouns, verbs, adjectives and adverbs, rather than 'grammatical words' such as prepositions (e.g. 'in,' 'at'), determiners (e.g. 'the,' 'a') or conjunctions (e.g. 'and,' 'or,' 'however') (Maton and Doran 2017). Table 5.3 summarizes the translation device for word-grouping.

In Figure 5.2, we show values according to the category of wording (whether a technical or everyday word) and the relative ESD strength (a technical word condensing more ESD than an everyday word). These values attributed to the wording allowed us to estimate the *range* of semantic density in classroom discourse in the Chemistry lessons as follows. Based on the

Table 5.2 Translation device for epistemic–semantic density of words

ESD	Type	Subtype	Sub-subtype	Points
+	technical	conglomerate	-properties	+8
			-elements	+7
		compact	-properties	+6
			-elements	+5
	everyday	consolidated	specialist	+4
			generalist	+3
		common	nuanced	+2
−			plain	+1

Adapted from Maton and Doran (2017).

Table 5.3 Word-grouping tool for epistemic–semantic density in English discourse

ESD	Type	Points
ESD ↑↑↑	embedded	+3
ESD ↑↑	categorized	+2
ESD ↑	located	+1

Aadapted from Maton and Doran (2017).

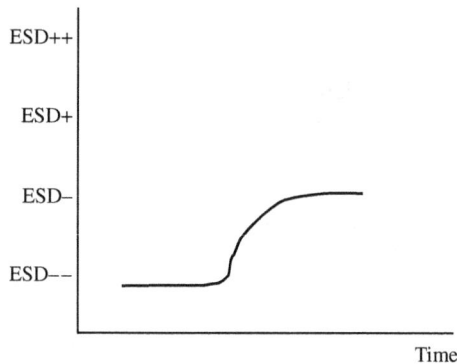

Figure 5.2 An upward shift or 'packing' movement in the classroom discourse.

measurement of lexical density (Halliday 2004), we divided the sum of points attributed to the content words by the number of content words in a clause. It is important to highlight that through this calculation, we are not analyzing the lexical density but creating a measure for the average strength of semantic density.

The sum of content word points in a clause included nouns, verbs, adjectives and adverbs but excluded auxiliary and linking verbs, prepositions, determiners and conjunctions. 'Proxy' words score according to the word or wording-group they refer to but score only as a single word. We also left out clauses that communicate class management.

Data source

Our study is in alignment with the ethical principles outlined by the Brazilian guidelines for research involving humans as the subjects of study and includes concern about informed consent and confidentiality. Student names are omitted, and teachers have been assigned pseudonyms. Our data comprises classroom discourse fragments from 12 lessons of grade 10 Chemistry by four teachers at different urban Brazilian schools in a town in the country-side of the state of Bahia. The lessons covered the topics of physical states of matter, atomic structure and pure substances and mixtures. In most, teachers taught through conventional methods and activities such as oral exposition near the blackboard and solving textbook exercises collectively with the class. The schools included public and private institutions in order to embrace the sociocultural diversity in our sample, though this was not an explored varia-ble in the analysis we present in this chapter. All teachers who taught the observed lessons have a degree in Chemistry and from five to over 20 years of professional experience as teachers. They did not get any special training to give these lessons, nor did we ask them to change their pedagogical style of teaching. Our involvement was to merely observe and register their les-sons with their permission.

We observed and recorded the audio from the four classrooms through-out a common teaching unit in Brazilian schools that lasts approximately two months. The audio recordings of all classes were transcribed verbatim. After this, the research team conducted an analysis in order to get an overview of the data. To select episodes, we made a detailed analysis of the semantic density degrees by applying our translation devices. Although we made the analysis from the Portuguese version of the transcripts, we decided to show only the English translation of them due to the lack of space.

Our analyses are illustrated with short episodes or fragments of those les-sons. We adopted a similar analysis method as Matruglio *et al.* (2013), in which we selected teaching episodes where we identified changes in the strength of ESD, then analyzed the language in those episodes using LCT. In all excerpts, the teachers' talk is marked T, and the students' talk is marked S. In the tran-scripts, some adaptations were made due to written language conventions.

Analysis

In the following transcript (episode #1) from the lesson by Carlos on physi-cal states of matter, we identified a variation from level ESD-- to level ESD-, strengthening the semantic density according to our first translation device, in a process of 'packing' meanings by moving the discourse from everyday to technical language through the use of scientific definitions. Indeed, in this variation, the teacher highlights that he is going to use a more 'elaborate' language to explain the same phenomenon a student had previously spoken about using everyday language. A profile to represent this episode is depicted in Figure 5.2, comprising an upward shift in ESD.

T: Like you said ... reviewing this knowledge she mentioned, ok, by slow she meant to say evaporation, imagine that you jump in a pool, obviously you get wet, right, but after a while you're dry. What took place, a process::... slow, like she said, but natural where that liquid changed to a, right? Boiling would be what?

S: A fast process.

T: Boiling, ok, in more elaborate language we'll say that there occurred an adequate temperature graduation for that substance to change physical state. Water, what's the boiling point, one hundred degrees, when you give a temperature to this substance, which in this case is one hundred degrees Celsius to change its physical state, it passes from a liquid state to a vapor state, so it's important to know that there are three changes of state, got it? ... From a gas to a liquid ... is the temperature there increasing or decreasing?

In episode #2, we have the same content in an extract from Bento, but now the initial variation is opposite that of episode #1, with a shift from level ESD– to level ESD––. In this case, the teacher first introduces the scientific concepts to then 'unpacking' their meanings in a known instance by using everyday language. According to Maton (2013), this downward shift in ESD (Figure 5.3) from more condensed ideas to simpler and concrete understanding is a very frequent profile of classroom discourse.

T: Boiling, but we can call it evaporation yeah, or vaporization right, so what happens, this vaporization can be, it can be called evaporation right, boiling or calefaction, that'll depend on how it passes, evaporation is a slow phase from a liquid to a gaseous state, boiling ... is a passage that's a little bit quicker, right, when the liquid's there bubbling, you know, it's a change of state:: liquid to a gaseous state in a faster way, calefaction is a sharp change, quickly, an example of this is when you have a totally hot surface, for example when you put a pot on the stove and forget it, oops I got to go take care of that problem there, so you put a little pot of water and what happens ... sheeshee ... right, it evaporates, yeah, right, so that's calefaction, but the opposite, the opposite can happen ok, so passing from a gaseous state to a liquid and from a liquid to a gas, these examples, or to a solid state, examples, give me an example then, gas to liquid, from liquid to a solid?

Regarding the range of ESD in these two episodes, our second translation device indicated a similar range of variation, although Carlos' discourse got a lower value than that of Bento. In Table 5.4, we compare the clauses with the lowest and the highest values in both episodes. The main source of the highest ESD in these clauses is the quantity of technical words. In the lowest strength ESD, most content words are everyday words, though Bento used more nuanced words like *abrupt, surface* and *flame* than Carlos did, which raised the ESD strength slightly when compared with Carlos' talk. We also included the values attributed to the content words and about how

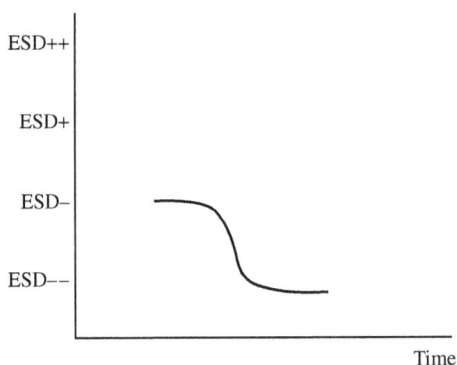

Figure 5.3 A downward shift.

Table 5.4 Highest and lowest strength of epistemic–semantic density in Carlos's and Bento's episodes.

Carlos
Boiling (5), ok, in more **elaborate** (2) **language** (2) we'll say that there **occurred** (2) an **adequate** (2) **temperature** (2) **graduation** (5) for that **substance** (5) to **change** (1) **physical** (5) **state** (5) *(highest strength ESD)* *(36 points for 11 content words = 3,27)*
imagine that **you** (1) **jump** (1) in a **pool** (1), obviously **you** (1) **get** (1) **wet** (1), right, but **after** (1) a **while** (1) **you**'re (1) **dry** (1). What took place, a **process** (3) *(lowest strength ESD)* *(14 points for 11 content words = 1,27)*
Bento
evaporation is a **slow phase** from a **liquid** to a **gaseous state**, **boiling** … is a passage that's a little **bit quicker**, right, **when** the **liquid**'s there **bubbling**, you know, it's a **change** of **state**:: **liquid** to a **gaseous state** in a **faster way** *(highest strength ESD)*
calefaction is a **sharp change**, **quickly**, an **example** of this is **when you have** a **totally hot surface**, for example **when you put** a **pot** on the **stove** and **forget** it *(lowest strength ESD)*

we calculated the results for Carlos' fragments by applying our second translation device to estimate the ESD strength in an illustrative manner of our analysis.

Nevertheless, these selected episodes did not exhibit the highest condensed meanings which Chemistry knowledge can reach. According to van Berkel *et al.* (2009), students can only develop an understanding of real-life phenomena in a mainstream Chemistry curriculum when they understand the corpuscular 'building-block' of atoms and molecules. Chittelborough (2014) has also argued that the study of Chemistry is essentially about the atomic theory of matter. The next episodes will present discourses covering knowledge involving the submicroscopic level of Chemistry. In episode #3, a lesson on physical states of matter, Durval explains the relationship between physical states, energy and inner organization of matter.

T: So, it presents three physical states: solid, liquid and gas, right, it's just that each physical state possesses a particularity, for example, in a solid state, the matter will be extremely organized, it has a fixed volume, ok? ... While in a liquid state, it'll be a little more disorganized and will have a variable volume. In the third, in the gaseous state, it'll have an extremely disorganized and an extremely variable volume, got it, ok? So, it possesses a fixed shape and volume, right, which is to say, the force of attraction between the molecules or atoms, um ... for the solid state, it's greater, as much for the force of attraction as for the force of repulsion, as they are very close, ok. But in the liquid state it possesses a variable shape and a fixed volume, ok? The attraction force can be (...) it's approximately equal to the repulsion force between the molecules, right? In the gaseous state, it will have a variable shape and volume, right, the forces of attraction and repulsion are far less between the constituent molecules of a gas, ok? So:: we can change the physical state modifying the temperature and pressure, increasing the temperature indicates what? Absorbing energy, reducing the temperature indicates what? Releasing energy from the matter, right? So, there you have a solid state to a liquid, liquid to a gas. The changes of physical state receive determined names. From a solid to a liquid, what'll be the name?

This episode starts with the discourse at level ESD–, then the teacher shifts to level ESD+ and keeps the strength of ESD around this level until the end of the episode. In his explanation, he employs entities such as molecules and atoms to teach the organization of matter and to explain properties at a macroscopic level. In his discourse, processes and events such as attraction and repulsion are nominalized as attraction and repulsion forces or are related to energy absorption and liberation. The profile of ESD for episode #3 is depicted in Figure 5.4 and assumes an upward shift or 'packing.'

In episode #4, Marina also appeals to the corpuscular theory to explain the properties of matter at a macroscopic level. Nevertheless, she starts at level ESD+ since her explanation involves molecules and the amount of

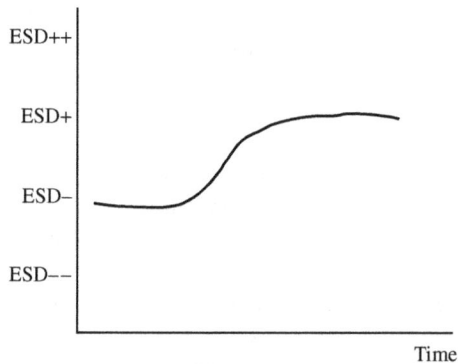

Figure 5.4 An upward semantic wave shift.

space between them to characterize the physical state of matter. After that, her discourse assumes a downward shift to level one, with the concrete example of milk in a glass and on a plate, in a movement of 'unpacking' the meanings.

T: Because of the heat, the temperature. This makes the molecules move away ... when the molecules move away from one another, there the movement gets better, doesn't it? With more space between them, so the movement gets better. Then what happens to this substance which is a liquid, it has a defined amount, which means, a defined volume, but it doesn't have a shape, how can I understand this? When I put milk there in the glass, what is going to happen to the milk? It takes the shape of the glass, if I take the milk and put it in a plate?

The semantic profile of this episode is represented by the semantic wave in Figure 5.5. The shape of this profile assumes a 'downward escalator' (Maton 2013). According to Maton (2013), this is a very frequent profile which represents a pedagogic practice where teachers 'unpack' and exemplify meaning with concrete understandings, as in the case of the shape of milk in a glass or a plate.

Contrasting episodes #3 and #4 with regard to our second translation device will reveal that, although both episodes include level three, their semantic ranges are quite different from each other. In Table 5.5, we present the highest and lowest ESD for clauses in the two episodes.

Marina's discourse, although involving molecules, does not reach the same ESD strength as that of Durval, even though he does not talk about particles. The calculated value for Marina's strongest ESD clause achieves the same value as this clause from Durval: 'So, it presents three physical states: solid, liquid and gas, right, it's just that each physical state possesses a particularity.' In the latter, there is no reference to molecules or atoms, which means that the calculated value for estimating the semantic range exhibits an overlap between levels two and three. This finding led us to ask the following

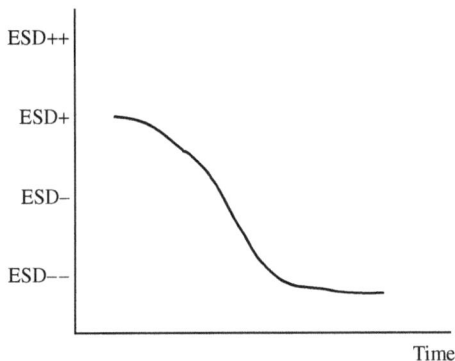

Figure 5.5 A 'downward escalator' profile.

Table 5.5 Stronger and weaker strength of epistemic–semantic density in Durval's and Marina's episodes.

Durval
So, it **possesses** a **fixed shape** and **volume**, right, which is to say, the **force of attraction** between the **molecules** or **atoms**, um … for the **solid state**, it's **greater**, as much for the **force of attraction** as for the **force of repulsion**, as they **are very close**, ok. (*stronger ESD*)
But in the **liquid state** it **possesses** a **variable shape** and a **fixed volume**, ok? (*weaker ESD*)

Marina
This makes the molecules move away … **when the molecules move away** from **one another**, there the **movement gets better**, doesn't it? (*stronger ESD*)
When I put milk there in the **glass**, what is going to **happen** to the **milk**? It **takes** the **shape of the glass**, if **I take** the **milk** and **put** it in a **plate**? *(weaker ESD)*

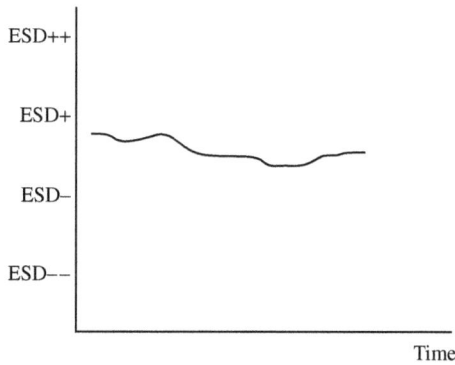

Figure 5.6 A 'flatline' profile.

question: is the strengthening of ESD a matter of knowledge in such a way that changes of the nature of knowledge can alter the strength of ESD or is the strengthening of ESD exclusively a matter of linguistics? The next two episodes perhaps expand our understanding related to this question.

In episode #5, Carlos characterizes the atom, drawing on some features of the electron. According to our first translation device, the discourse in this fragment keeps a relatively high ESD strength due to the use of particle theory in his explanation. The wording tool revealed little variation in the range of ESD since all clauses include technical words like *electrosphere*, *cathode rays* and *acceleration*. A profile of ESD in this episode is depicted in Figure 5.6 and represents a high flatline which reflects the use of abstract and condensed concepts (Maton 2013).

T: It was proven that the electron has mass, a minimal mass … but it exists, ok? So, not every atomic mass is found in the nucleus, the electrons that are located in the electrosphere also have … mass, it's a minimal mass, but there's mass, ok? This was proven in the time in which the electrons were called cathode rays, when a beam of light, which represented these rays which today are considered electrons, were able to move a structure and moved this structured … it's because there exists a force and the force is which? Mass times acceleration, so there's force time mass, then these cathode ray beams were able to move a structure and it was discovered that the electron also had mass, so, the goal is that you understand that not all atomic mass is located in the nucleus, got it?!

Marina also characterizes atomic structure, but she uses an analogy to teach it, as portrayed in episode #6. This weakens the ESD strength from the level ESD+ to ESD-- at the beginning, but she raises the strength towards the end of the episode from level ESD-- to ESD+.

T: If an atom were the size of a football stadium, the nucleus would be the size of a ladybug in the middle of the field, so imagine that the atom, it's invisible, the atom, it's invisible, ok? So, imagine the size that the nucleus would be, being invisible, can you get the picture? Because if the atom were the size of a football stadium, the nucleus would be the size of a ladybug, of a beetle, oh. But that's if it were the size of a?

S: Stadium

T: Imagine that it is, it's so small that we can't even see it. So, imagine that the nucleus, it's even small than that.

S: Smaller?

T: Imagine Maracanã, the Maracanã stadium, if you take the whole stadium, not just the field, and you look, you'll see that it makes a circle and it's round like this, now imagine a marble there in the middle where the player rolls the ball to start the game in the middle of the stadium, this marble would be the nucleus of the atom and all around would be the electrosphere where the electrons are, ok, that is, the electrosphere around is where the electrons are and the nucleus is there in the middle which is the marble where the protons and neutrons are.

According to this, the profile of this episode assumes a shape that may promote knowledge-building (Maton 2013) as depicted in Figure 5.7. This wave of ESD represents the 'unpacking' and 'packing' movements in the classroom discourse.

Applying the wording tool to episode #6, however, disclosed a richer variation that our first translation device does not capture in its entirety. With this tool, we noticed that even when the teacher is talking about the universe of particles, when we would expect a higher ESD level, mixing it up with everyday words can weaken the ESD strength, which was assisted by the use of an analogy. This is illustrated by the first clause of this episode: '*If an atom were the size of a football stadium, the nucleus would be the size of a ladybug in the middle of the field.*' The introduction of an analogy allows variation in the ESD strength, thus avoiding the flat line we observed in Carlos' talk. Table 5.6

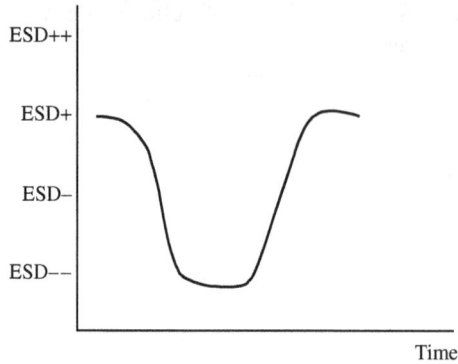

Figure 5.7 A downward shift followed by an upward shift.

Table 5.6 Stronger and weaker strength of epistemic–semantic density in Carlos's and Marina's episodes

Carlos
This was **proven** in the **time** in which the **electrons** were **called cathode rays,** when a **beam** of **light**, which **represented** these **rays** which **today** are **considered electrons**, were **able** to **move** a **structure** and **moved** this **structure** … it's because there **exists** a **force** and the **force** is which? *(strongest ESD)*
So, the goal is that you understand that not **all atomic mass** is **located** in the **nucleus**, got it?! *(weakest ESD)*
Marina
this **marble** would be the **nucleus** of the **atom** and **all** around would be the **electrosphere** where the **electrons** are *(strongest ESD)*
now **imagine** a **marble** there in the **middle** where the **player rolls** the **ball** to **start** the **game** in the **middle** of the **stadium** *(weakest ESD)*

displays the highest and lowest ESD for clauses in these two episodes. Marina's clause with the strongest ESD is comparable to Carlos' weakest ESD clause.

Both Durval in episode #3 and Carlos in episode #5 raise the ESD strength by including many concepts or many technical words in their clauses when they teach about the universe of particles. Marina addresses the same content from the opposite direction: she dilutes the technical words and concepts through clauses rich in everyday words. While Durval explains the inner organization of matter as a balance between forces, Marina prefers to use a metaphor of movement and space ('when the molecules move away from one another, there the movement gets better, doesn't it?').

Discussion

School science as a hybrid discourse

Under a perspective based on language studies, school Science discourse was characterized to be a *hybrid discourse practice* (Kamberelis and Wehunt 2012) or an *interlanguage* – a specific discourse that involves the mixing together of scientific expressions with everyday language (Nygård Larrson and Jakobsson 2020; Lemke 1990). In the teaching practices of Science classrooms, the significance and meanings of technical words and expressions 'are often discussed and explained in an everyday perspective and language' (Nygård Larrson and Jakobsson 2020). This is probably why teachers' discourse hardly ever strengthens to the highest ESD levels: even when teachers' discourses shift towards higher levels, mixing the discourse mitigates that strengthening.

This appears to be the option Marina chooses to teach the abstract content of particle theory to her students, through the use of analogy and metaphors. The analysis by the wording tool clearly points out how the use of a hybrid discourse weakens ESD, where we would expect a higher degree of ESD if we analyzed the same episode with our first translation device. In this sense, Maton warns us that weakening ESD of educational knowledge to 'unpack' the scientific concepts is not intrinsically negative since it is by 'translating' a technical term into common-sense understandings reduces its range of meanings, which is the purpose: to provide a point of entry for novices into those meanings' (Maton 2013: 15).

The analogical reasoning is based on a semantic relationship between two domains: a base domain and a target domain (Vosniadou and Ortony 1989). According to Poulet (2016), the use of analogical reasoning tools may strengthen semantic density if the involved knowledge is integrated with tools of abstraction like metaphors. Nevertheless, in Chemistry classroom discourse, we notice an ESD weakening through the use of analogy, as in the case of episode #6 with Marina. She uses concrete knowledge as the base domain in her recontextualization of atomic structure theory – the target domain – and it decreases the ESD strength through the introduction of everyday words alongside technical ones, such as electrons, nucleus and electrosphere. When taught by Carlos in episode #5, the same content remained at a higher ESD level, resulting in a semantic wave represented as a flat line.

The use of hybrid language in Science teaching is supposed to facilitate the acquisition of more scientific language by students and, in this sense, it plays a crucial role to engage students in advanced Science learning (Kamberelis and Wehunt 2012; Lemke 1990). Because of this, such a hybrid discursive practice is considered favourable for learning, as well as for a more inclusive teaching practice (Hanrahan 2006). On the other hand, it is not uncommon that much of the usage of this particular language takes place in an implicit way in teachers' discourse, which may cause confusion and misunderstandings between students, as the appropriate context in which the words and expressions belong does not become clear to them (Nygård Larrson and Jakobsson 2020; Jakobsson and Serder 2016; Fang 2005). In this sense, it is worth pointing out that many everyday words like substance, mass, force, temperature and volume have specific definitions and precise uses in Science that differ from their use in an everyday context. Moreover, Halliday and

Martin (1993) warn us that building meaning in scientific discourse is underpinned not only by the use of technical words and expressions but also on the taxonomies in which they are situated.

Limitations of our translation device

There is a form of metadiscourse in teaching in which the subject of talk is a reflection of the Chemistry knowledge to be studied and learned. It is an unanticipated type of discourse for our device to analyze the ESD level. It can include:

a) *Didactical reflection on the content taught.* In the following excerpt, Marina explains to her students her choices about the atomic model she will teach them, telling the main reason to avoid one of the models:

Because like this, guys, the atom that we'll study, that we'll use, which is the most used didactically, is the union of Rutherford's model with Bohr's model, because this last model here from Schrödinger, we can't study the atom behaving like a cloud, speaking of the cloud, it's a more modern model, but when you'll talk about, how are you going to study a cloud, right? Which is not a defined thing, so we can't study it, so the model we'll study is the combination of Rutherford's model with Bohr's model.

b) *An epistemological reflection on the scientific knowledge,* as Bento teaches his students when he is talking about the evolution of the idea of the atom:

Some scientists who were appearing at the time, who were observing, who were questioning, who weren't completely wrong, the atoms of other previous scientists, but they were improving, perfecting until the idea arrived of the atom there is today, ok, so they all contributed, from back with the philosophy of the first thought, the first observation, the first idea about matter, the composition of matter served to arrive at the science of today, to get to Chemistry as it is today, right. So, it was important, all of these people there who passed through all of these periods, thinking, questioning, formulating hypotheses, experimenting to get to the idea of today.

As both discourses transmit values and judgement, they are forms of *axiological condensation* of affective and aesthetic stances, and they require an additional device to capture *axiological-semantic density* (Maton and Doran 2017, Maton 2014a). Besides these examples, Mortimer and Scott (2003) also warn us about a third mode of discourse in Science classes which involves management and organizational issues, another kind of content that is not included in our translation devices. They also alert us to

Initiation-Response-Evaluation sequences, repetition, emphasis and so on, which have a role in classroom discourse, something that we do not have a way of expressing through this use of LCT.

Conclusion

The translation devices we introduced here have the power to complement each other. They also have some limitations. The one that shows variations of the level of Chemistry knowledge helps us realize that teaching discourse can meet or depart from Johnstone's claim, which argues that Chemistry 'should be presented in a way that capitalizes on what students are familiar with' and that concepts 'must be built from the macroscopic and gradually be enriched with submicroscopic and representational aspects' (Johnstone 2010: 28). Through this translation, we can analyze and evaluate the choices teachers make during instruction, taking into account the nature of Chemistry knowledge.

On the other hand, the wording tool gives us a finer estimation of the semantic range between the highest and lowest strengths, and can be viewed as a means to calibrate the semantic scale with greater precision (Maton 2013). This way, we can also analyze and evaluate the efficacy of the use of hybrid discourse in Chemistry teaching, with a sense of estimating if a particular discourse delivers the scientific concepts and allows students to learn the Science discourse in its entirety. According to Aikenhead Science instruction in schools is a kind of cross-cultural event for most students:

> If students are going to cross the border between everyday subcultures and the subculture of science, border crossings must be explicit and students need some way of signifying to themselves and others which subculture they are talking in, at any given moment.
>
> (Aikenhead 1996: 30)

Concerning the limitations, we can say that Chemistry classroom discourse is more than simply a discourse on the content of Chemistry. It involves Initiation-Response-Evaluation sequences, repetition, emphasis and several types of discourse – for example, one of management and organizational issues – that are beyond the scope of the analysis centred in epistemic–semantic density.

The use of these translation devices to build semantic waves also relates to the dispositions of the actors, as Maton (2013) says. Research on semantic profiles is exploring practices in education, not exclusively on classroom discourse and curriculum, but including teacher education as well. It offers great potential to help teachers build their teaching sequences. It is important to find ways to bring the new knowledge LCT provides to the educational field to Chemistry teachers.

References

Aikenhead, G.S. (1996) 'Science education: Border crossing into the subculture of science' *Studies in Science Education*, 27(1): 1–52.

Becker, N., Stanford, C., Towns, M. and Cole, R. (2015) 'Translating across macroscopic, submicroscopic, and symbolic levels: The role of instructor facilitation in an inquiry-oriented physical Chemistry class,' *Chemical Education Research and Practice*, 16(4): 769–785.

Blackie, M.A.L. (2014) 'Creating semantic waves: Using Legitimation Code Theory as a tool to aid the teaching of Chemistry,' *Chemical Education Research and Practice*, 15(4): 462–469.

Chittelborough, G. (2014) 'The development of theoretical frameworks for understanding the learning of Chemistry,' in I. Devetak, and S. A. Glažar (eds) *Learning with understanding in the Chemistry classroom*, Springer.

Danielsson, K., Löfgren, R. and Pettersson, A.J. (2018) 'Gains and losses: Metaphors in Chemistry classrooms,' in K. Tang and K. Danielsson (eds) *Global developments in literacy research for science education*, Springer.

Fang, Z. (2005) 'Scientific literacy: A systemic functional linguistics perspective,' *Science Education*, 89(2): 335–347.

Freire, P. (2018) *Pedagogy of the oppressed* (4th edition), Bloomsbury.

Fensham, P.J. (2004) *Defining an identity. The evolution of science education as a field of research*, Kluwer.

Georgiou, H. (2016) 'Putting physics knowledge in the hot seat: The semantics of student understandings of thermodynamics,' in K. Maton, S. Hood, and S. Shay (eds) *Knowledge-building: Educational studies in Legitimation Code Theory*, Routledge.

Gilbert, J.K. and Treagust, D. (2009). 'Towards a coherent model for macro, submicro and symbolic representations in chemical education,' in J.K. Gilbert and D. Treagust (eds) *Multiple representations in chemical education*, Springer.

Halliday, M.A.K. (2004) *Introduction to functional grammar* (4th edition), Routledge.

Halliday, M.A.K. and Martin, J.R. (1993) *Writing science: Literacy and discursive power*, Taylor and Francis.

Hanrahan, M.U. (2006) 'Highlighting hybridity: A critical discourse analysis of teacher talk in science classrooms' *Science Education*, 89(1): 8–43.

Jakobsson, A. and Serder, M. (2016). 'Language games and meaning as used in students encounters with scientific literacy test items,' *Science Education*, 100(2): 321–343.

Johnstone, A.H. (1982) 'The nature of Chemistry,' *School Science Review*, 64: 377–379.

Johnstone, A.H. (1993) 'The development of Chemistry teaching,' *Journal of Chemical Education*, 70(9): 701–705.

Johnstone, A.H. (2000) 'Teaching of Chemistry – logical or psychological?' *Chemistry Education: Research and Practice in Europe*, 1(1): 9–15.

Johnstone, A.H. (2010) 'You can´t get there from here,' *Journal of Chemical Education*, 87(1): 22–29.

Kamberelis, G. and Wehunt, M.D. (2012) 'Hybrid discourse practice and science learning,' *Cultural Studies of Science Education*, 7: 505–534.

Lemke, J.L. (1990) *Talking science: Language, learning, and values*, Ablex.

Liu, Y. and Taber, K.S. (2016) 'Analyzing symbolic expressions in secondary school Chemistry: Their functions and implications for pedagogy,' *Chemistry Education Research and Practice*, 17(3): 439–451.

Martin, J.R. (2013) 'Embedded literacy: Knowledge as meaning' *Linguistics and Education*, 24(1): 23–37.

Maton, K. (2013) 'Making semantic waves: A key to cumulative knowledge-building. *Linguistics and Education*, 24(1): 8–22.

Maton, K. (2014a) *Knowledge and knowers: Towards a realist sociology of education*, Routledge.

Maton, K. (2014b) 'Building powerful knowledge: The significance of semantic waves,' in B. Barrett and Rata, E. (eds) *Knowledge and the future of curriculum. International studies in social realism*, Palgrave.

Maton, K. (2020) 'Semantic waves: Context, complexity and academic discourse,' in Martin, J. R., Maton, K. and Doran, Y. J. (eds) *Accessing academic discourse: Systemic functional linguistics and Legitimation Code Theory*, Routledge.

Maton, K. and Chen, R.T-H. (2016) 'LCT in qualitative research: Creating a translation device for studying constructivist pedagogy,' in K. Maton, S. Hood and S. Shay (eds) *Knowledge-building: Educational studies in Legitimation Code Theory*, Routledge.

Maton, K. and Doran, Y.J. (2017) 'Semantic density: A translation device for revealing complexity of knowledge practices in discourse, part 1 – wording.' *Onomázein*, (1): 46–76.

Matruglio, E., Maton, K. and Martin, J.R. (2013) 'Time travel: The role of temporality in enabling semantic waves in secondary school teaching,' *Linguistics and Education*, 24(1): 38–49.

Mortimer, E.F. and Scott, P.H. (2003) *Meaning making in secondary science classrooms*, Open University Press.

Nakhleh, M.B. (1993) 'Are our students conceptual thinkers or algorithmic problem solvers?' *Journal of Chemical Education*, 70(1): 52–55.

Nelson, P.G. (2002) 'Teaching Chemistry progressively: From substances, to atoms and molecules, to electrons to nuclei,' *Chemistry Education: Research and Practice in Europe*, 3(2): 215–228.

Nygård Larrson, P. and Jakobsson, A. (2020) 'Meaning-making in science from the perspective of students' hybrid language use,' *International Journal of Science and Mathematics Education*, 18(5): 811–830.

Poulet, C. (2016) 'Knowledge and knowers in tacit pedagogic contexts: Free mansonry in France,' in K. Maton, S. Hood, and S. Shay (eds) *Knowledge-building: Educational studies in Legitimation Code Theory*, Routledge.

Rootman-le Grange, I. and Blackie, M.A.L. (2018) 'Assessing assessment: In pursuit of meaningful learning,' *Chemical Education Research and Practice*, 19(2): 484–490.

Santos, B.F. and Mortimer, E.F. (2019) 'Ondas semânticas e a dimensão epistêmica do discurso na sala de aula de Química,' *Investigações sobre o Ensino de Ciências*, 24(1): 62–80.

Scott, P.; Mortimer, E.F. and Ametller, J. (2011) 'Pedagogical-link making: A fundamental aspect of teaching and learning scientific conceptual knowledge,' *Studies in Science Education*, 47(1): 3–36.

Sjöström, J. and Talanquer, V. (2014) 'Humanizing chemical education: From simple contextualization to multifaceted problematization,' *Journal of Chemical Education*, 91(8): 1125–1131.

Stieff, M., Ryu, M. and Yip, J.C. (2013) 'Speak across levels – generating and addressing levels of confusion in discourse,' *Chemical Education Research and Practice*, 14(4): 376–389.

Stuckey, M., Hofstein, A., Mamlok-Naaman, R. and Eilks, I. (2013) 'The meaning of 'relevance' in science education and its implications for the science curriculum,' *Studies in Science Education*, 49(1): 1–34.

Talanquer, V. (2011) 'Macro, submicro and symbolic: The many facets of the Chemistry "triplet",' *International Journal of Science Education*, 33(2): 179–195.

van Berkel, B., Pilot, A. and Bulte, A.M.W. (2009) 'Micro-macro thinking in chemical education: Why and how to escape,' in: J.K. Gilbert, and D. Treagust (eds) *Multiple representations in chemical education*. Springer.

Vosniadou, S. and Ortony, A. (1989) *Similarity and analogical reasoning*. Cambridge University Press.

Wu, H. (2003) 'Linking the microscopic view of Chemistry to real-life experiences: Intertextuality in a high-school science classroom,' *Science Education*, 87(6): 868–891.

6 Radiation Physics in theory and practice

Using Specialization to understand 'threshold concepts'

Lizel Hudson, Penelope Engel-Hills, and Chris Winberg

Introduction to radiation physics and radiation therapy

Radiation Physics is the study of ionizing electromagnetic radiation, including γ-rays obtained by the decay of an atomic nucleus and X-rays produced when electrons strike a target. Radiation Physics is the science that underpins the practice of Radiation Therapy. Radiation therapists treat patients with cancer, as part of a multidisciplinary oncology team. Advances in cancer management have been influenced by technological advances in medical imaging (Baumann *et al.* 2016) and Radiation Therapy equipment. Another major impact has been the expansion of radiation therapists' scope of practice to include decision-making in patient management (Harnett *et al.* 2014; Harnett *et al.* 2018). These developments in the profession have resulted in different approaches to the training of radiation therapists, in order to prepare them adequately for the changing clinical environment, as well as for their expanding roles. It is clear that students need scientific knowledge to enable new and evolving forms of practice. That is why consideration of key concepts of Radiation Physics is important in the education of radiation therapists, who face unknown future contexts.

Many lecturers and researchers have found the idea of 'threshold concepts' (Meyer and Land 2003, 2005) to be useful, but also confusing. For this reason, this study aimed to address the research question: How could threshold concepts in Radiation Physics be described in an empirically grounded and theoretically consistent way for the benefit of lecturers, students and clinical educators? This chapter presents a way of unpacking and tackling threshold concepts in Radiation Therapy education using concepts from the Specialization dimension of Legitimation Code Theory to demonstrate the development of what is termed an *élite code* orientation over time.

Brief overview of the literature on threshold concepts

The idea of threshold concepts emerged from an educational study into the disciplinary characteristics of the field of economics. Meyer and Land (2003) noted that 'certain concepts were held, by economists, to be central to the mastery of their subject.' They described these concepts as 'threshold'

DOI: 10.4324/9781003055549-8

because they were analogous to a doorway into the discipline. In subsequent studies (e.g. Meyer and Land 2005; Meyer *et al.* 2006) researchers linked students' mastery of disciplinary knowledge to their understanding of its threshold concepts. The Threshold Concept Framework clusters together a range of ideas about why students might experience difficulty in mastering complex disciplinary knowledge and, in the context of this study, why applying them successfully in the clinical environment is so challenging. Cousin (2006) notes that all disciplines have threshold concepts that are fundamental to that discipline, for example, a limit in Mathematics (Scheja and Pettersson 2010) and atomic structure in Physics (Park and Light 2009). Understanding why students experience difficulties is the first step towards supporting them. In this regard, the Threshold Concept Framework provides a list of characteristics that lecturers can explore when modifying or redesigning curricula (Dunn 2019).

Threshold concepts are distinguished from other concepts by their complexity, their high level of abstraction, and their centrality to the discipline. Threshold concept descriptors, as explained by Cousin (2010: 1–2) and Meyer and Land (2005) have particular key features:

1. Transformative: New understandings are 'assimilated into our biography,' becoming a part of 'what we know' and 'who we are.'
2. Irreversible: Although difficult to grasp, once a threshold concept is understood, the student is unlikely to forget it.
3. Integrative: Threshold concepts tend to integrate prior disciplinary concepts, thus mastering a threshold concept can enable the student to make connections across the curriculum. 'Things start to click into place.'
4. Bounded: Threshold concepts occur in disciplinary knowledge; they are not part of everyday knowledge or common sense.
5. Troublesome: Threshold concepts are 'troublesome' because they are complex and challenging and, to a novice, seem 'counter-intuitive, alien or seemingly incoherent.'
6. Discursive: The idea that threshold concepts are associated with disciplinary discourses was a later addition to the framework.

Meyer and Land (2003) included the concepts of 'liminal spaces' and 'states of liminality' to explain the process of learning a threshold concept. Cousin argues that while most learning involves recursive processes, in the case of threshold concepts, learning 'involves a strong emotional dimension concerning the student's identification with both the subject and his [sic] perceived capabilities' (Cousin 2010: 3). Zaky (2018) points out that liminal spaces and states are not static but dynamic and argues that teaching threshold concepts requires locating students' progress along a liminal continuum. The pre-liminal space represents an initial encounter with 'troublesome knowledge.' In the liminal space, the student undergoes recursive processes of integration and discarding prior understandings, which include concomitant ontological and epistemic shifts. Finally, successful students emerge in a post-liminal state of transformation and irreversibility. This process is illustrated in Figure 6.1.

Zaky (2018) re-organized the Threshold Concept Framework for the purpose of understanding the processes of learning threshold concepts. Our purpose in this study was to understand the characteristics of the knowledge implied by the Threshold Concept Framework, with specific reference to Radiation Physics.

While there is very limited literature on threshold concepts specific to Radiation Physics, a number of studies have identified threshold concepts in general Physics that have relevance to Radiation Physics. For example, 'probability' and 'energy quantization' were identified as threshold concepts for understanding atomic structure (Park and Light 2009), while 'electronic transition' and 'photon energy' were identified as threshold concepts for students' scientific understanding of atomic spectra (Körhasan and Wang 2016). These general Physics concepts were identified as threshold concepts because of their importance in enabling progression towards more advanced concepts. However, it is argued that transferring general Physics concepts to more specialized fields of study (e.g. Biophysics) is not helpful for identifying threshold concepts specific to these fields (Wolfson *et al.* 2014). In the case of Biophysics, Wolfson *et al.* (2014) point out that the interdisciplinary nature of threshold concepts in Biophysics means that they have characteristics that are distinct from pure Physics. Radiation Physics is an applied discipline in which pure Physics concepts are applied to the treatment of patients. Thus Radiation Physics contains threshold concepts that are not found in pure Physics, such as the isocentre (or centre of rotation) and the inverse

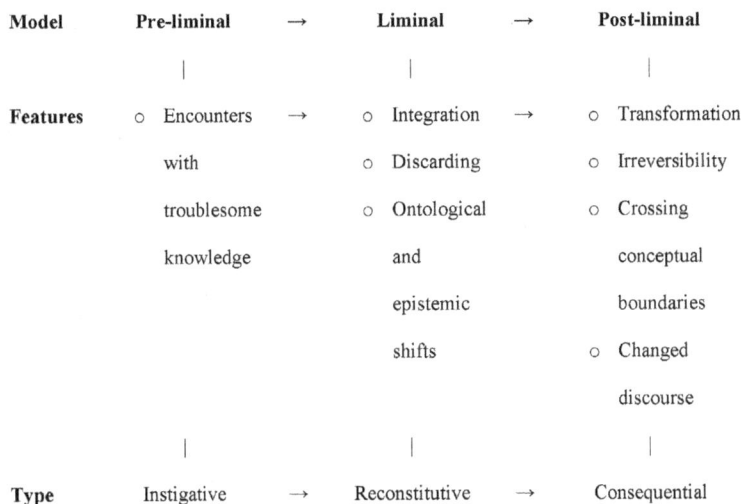

Model	Pre-liminal	→	Liminal	→	Post-liminal
Features	o Encounters with troublesome knowledge	→	o Integration o Discarding o Ontological and epistemic shifts	→	o Transformation o Irreversibility o Crossing conceptual boundaries o Changed discourse
Type	Instigative	→	Reconstitutive	→	Consequential

Figure 6.1 A relational view of the features of threshold concepts (Zaky 2018: 110).

square law of exponential radiation absorption used in radiation protection (Hudson *et al.* 2018).

Threshold concepts in professional education

In professional education, threshold concepts encapsulate the essential subject knowledge of the course of study that underpins professional practice (Baillie *et al.* 2013). Thus competent practice has been associated with mastery of threshold concepts in the disciplines associated with particular fields of practice (Dunn 2019). Much of the literature on threshold concepts in the health sciences relates to concepts underpinning care (Neve *et al.* 2017; Clouder 2005), general professionalism (Kinchin *et al.* 2010), or concepts in the disciplines that are common across health professions, such as Anatomy and Physiology (Weurlander *et al.* 2016). Inter-professionalism has also emerged as a threshold concept for inter-professional education and practice (Royeen *et al.* 2010).

Land (2011) proposes that if students in professional programmes fail to master threshold concepts, they will only be able to perform in a 'ritualized manner.' Wheelahan argues that 'students need to be inducted into disciplinary systems of knowledge, so they have access to the criteria used to judge knowledge claims, and over time, [and to] change the terms of the debate' (2015: 760). Recently, Fredholm *et al.* (2019) pointed out that practical experiences in the clinical environment have a similar effect to threshold concepts; that is, they transform thinking and identity and serve 'as a trigger for transformational learning, therefore making the discussion about 'practical thresholds' or thresholds in practice possible' (Fredholm *et al.* 2019: 2).

Critique of the Threshold Concept Framework

The Threshold Concept Framework has been debated in the literature, and its theoretical inconsistencies have been pointed out (e.g. Barradell 2013). Researchers have shown that the terms used to describe the characteristics of threshold concepts are often subjective and difficult to measure (Nicola-Richmond *et al.* 2018). Rowbottom (2007) claims that thresholds are 'unidentifiable,' while Walker (2013) suggests that the framework is a cognitive framework, rather than a framework for describing concepts. These critiques of the Threshold Concept Framework do not imply that the framework is not useful, but that the framework might need to be strengthened, in particular to avoid a conflation of knowers, those who are doing the learning, and knowledge, that which is being learned.

Theoretical framework: LCT Specialization

Legitimation Code Theory (LCT) offers many tools for analysis of knowledge practices. In this study, the dimension of Specialization (Maton 2014, Maton and Chen 2020) was drawn on to analyze how threshold concepts in

Radiation Physics were enacted in the curriculum and in pedagogies towards competent and safe clinical practice. Maton explains Specialization in terms of *epistemic relations* to objects and *social relations* to subjects (2014: 29). It is important to note that Specialization codes are referred to in relational terms, on continua of strengths of the two relations, rather than as typologies. On the *Specialization plane*, the *x*-axis represents social relations, and the *y*-axis represents epistemic relations. A disciplinary field, or a curriculum, or pedagogies or any form of practice can be located on the Specialization plane to reveal their relative strengths and weaknesses of *epistemic relations* (to other knowledge and the object of study) and *social relations* (to ways of knowing or knowers). Figure 6.2 is a graphical representation of the Specialization plane.

The four principal modalities created by the intersection of the two continua in Figure 6.2 are described by Maton (2016: 13) as follows:

- *knowledge codes* (ER+, SR−), where possession of specialized knowledge, principles or procedures concerning specific objects of study is emphasized as the basis of achievement, and the attributes of actors are downplayed;
- *knower codes* (ER−, SR+), where specialized knowledge and objects are downplayed and the attributes of actors are emphasized as measures of achievement;
- *élite codes* (ER+, SR+), where legitimacy is based on both possessing specialist knowledge and being the right kind of knower; and
- *relativist codes* (ER−, SR−), where legitimacy is determined by neither specialist knowledge nor knower attributes – 'anything goes.'

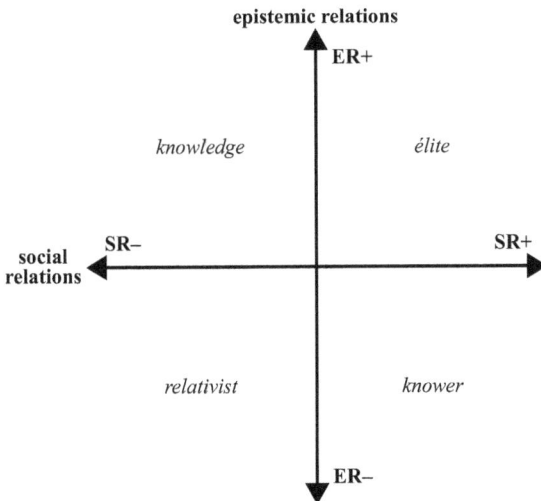

Figure 6.2 The specialization plane (Maton 2014: 30).

Maton (2016: 13) describes these codes as emphasizing 'what you know' (knowledge code), 'the kind of knower or practitioner you are' (knower code), both specialist knowledge and being a particular kind of knower (élite code) and emphasizing neither (relativist code).

Specialization codes were considered to be appropriate for analyzing how threshold concepts in Radiation Physics were enacted in theory-based learning and in clinical practice. Specialization affords a focus on epistemic relations to knowledge as well as social relations of 'practitioners' of the discipline. The use of Specialization in this study provided insights into threshold concepts in Radiation Physics, as well as their role in underpinning Radiation Therapy practice.

For the purpose of this study, the four codes on the Specialization plane (Figure 6.2) were adapted as in Figure 6.3.

Radiation Physics is the physical science that underpins Radiation Therapy practice. Radiation Physics can be characterized as a knowledge code, emphasizing epistemic relations (ER+) to knowledge and downplaying social relations (SR–) to knowers. In other words, it is an abstract scientific discipline. We thus expect to find threshold concepts that are abstract and complex in Radiation Physics. Radiation Physics underpins Radiation Therapy. Radiation Therapy is a clinical practice and has stronger social relations (SR+), but because it is underpinned by Radiation Physics, it also has stronger epistemic relations (ER+). For these reasons, it is described here as an élite code (ER+, SR+). Patient care is an ethical position, a mandated code of conduct for radiation therapists and a core competence for students. Although patient care requires underpinning by scientific knowledge, much patient care, such as attending to the comfort and well-being of the patient, has weaker epistemic relations and stronger social relations because it is dependent on

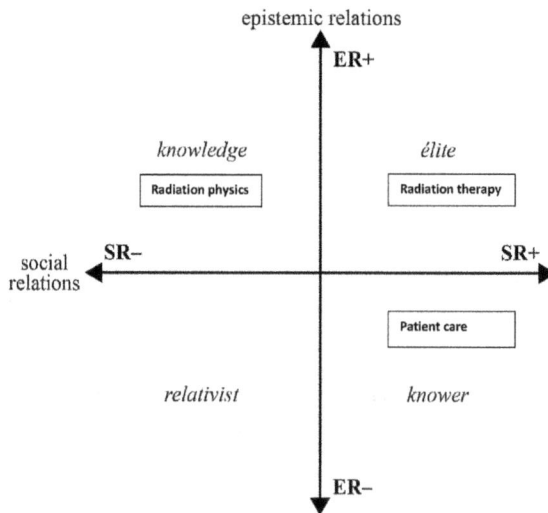

Figure 6.3 The specialization plane for Radiation physics.

appropriate dispositions. Thus patient care is a knower code (ER–, SR+). Relativist codes (weaker epistemic relations to knowledge and weaker social relations to practice) have no official space in the training or practice of radiation therapists.

Research design and methods

This study focused on the first-year Radiation Physics subject and addressed the research question: how could threshold concepts in Radiation Physics be described in an empirically grounded and theoretically consistent way for the benefit of lecturers, students and clinical educators? The research question called for an understanding of how threshold concepts in Radiation Physics were understood by lecturers, students and clinical educators. Lecturers are subject experts and experienced in teaching key concepts in a discipline; they, therefore, played an important role in the identification of threshold concepts. In this study, clinical educators were also included as experts because they understood the value of Radiation Physics in practice. Meyer and Land (2005) point out that because the experts have moved beyond threshold concepts, they find it difficult to identify concepts that they have long internalized. Thus, to ensure the accurate identification of threshold concepts, there is a need for a partnership between experts, educational researchers and students. Cousin calls this partnership a 'transactional curriculum inquiry' (2009: 202).

Participants' descriptions of first-year Radiation Physics were expected to be dependent on their contexts. Understanding was therefore anticipated as being determined by whether they were lecturers of first-year Radiation Physics, first-year students learning the subject, senior students reflecting on their learning in their first year or practising radiation therapists (referred to as clinical educators in the study).

The site selected to conduct this inquiry was the only university where a Bachelor of Science in Radiation Therapy was offered in South Africa. This study was approved by the Faculty Research Ethics Committee and permission was given by the relevant department to interview lecturers and students. Permission was also granted by clinical sites, where the clinical educators were interviewed.

Participants' perspectives on radiation physics knowledge

This section outlines the key issues raised by the participants in the 'transactional curriculum inquiry' (Cousin 2009). Data provided by participants provided different perspectives on what makes Radiation Physics challenging to learn, and challenging to teach.

Students' perspectives on first-year radiation physics

The students consistently described Radiation Physics as complex and difficult to understand. Reflecting on her first-year experience, a senior student comments: 'I honestly didn't understand a single thing' (Third-year student 3). A large part of the difficulties associated with Radiation Physics had to do with its abstract nature; first-year students used words like 'up there' (First-year student 1) or 'in the air' (First-year student 6) to describe their difficulty with the subject: '[The Physics lecturer is] like very up there … clever with Physics and I'm like … don't understand' (First-year student 3); 'But in Physics, I always feel it's – out of the air just here' (First-year student 6). For one interviewee, Radiation Physics was simply 'way too Physics-full' (First-year student 4).

Lecturers' perspectives on first-year radiation physics

The lecturers, who were either physicists or radiation therapists, did not perceive Radiation Physics as difficult. The physicists understood Radiation Physics as an abstract discipline, and they wanted students to achieve a level of abstract comprehension. The radiation therapists, on the other hand, did not experience Radiation Physics as particularly abstract. Although they described Radiation Physics as a discipline, they also recognized it as an integral part of Radiation Therapy practice. The physicists described Radiation Physics in the specialized language of the discipline, while the radiation therapists understood it in the language of Radiation Therapy practice. Both sets of lecturers interviewed described Radiation Physics as a blend of Physics and Therapy concepts: the 'concept of the x, y and z axis,' 'bending magnets,' 'waveguides,' 'anodes,' 'isocentre,' 'collimation,' 'virtual wedges,' and 'head of the machine' (Lecturer 4). They also understood the importance of the Radiation Physics concepts in underpinning skilled and safe practice: 'It's a high stakes environment. You know, if we conceptually get it wrong here, you know, you can imagine what the implication could be in clinical' (Lecturer 3).

For the physicists, Radiation Physics was separate from Radiation Therapy and worthy of study as a discipline in its own right that taught 'the process of thinking' (Lecturer 2) as much as the content of Physics. However, the physicists also understood Radiation Physics and Radiation Therapy as almost interchangeable:

> Radiation, how do we protect ourselves from it…? How do we utilize it to our maximum … capabilities … high dose to the tumour and then less dose to the surrounding tissue? That's the aim of Radiation Therapy and with Radiation Physics, we can understand that concept.
>
> (Lecturer 2)

In some cases, Radiation Physics was understood as a discipline with its own characteristics and properties – 'It is what it is' (Lecturer 1), but in most cases it was understood in relation to Radiation Therapy. As Lecturer 1 explains: 'I teach in a way that I learned how to set up in the department.' She explained that this was 'not necessarily an academic way of teaching,' but her teaching followed the sequence of practice:

What do you need? You need A to get to B and then from B, we can move to C and so that's how ... we need to straighten our patient. We need to look at the x, y and z. It's a three-point set up and that's where our ... set up starts. So it starts at straightening your patient and then choosing your reference and then from the reference moving to your isocentre and once your isocentre is there, we move onto the next step which is then the verification step. So that's how I sort of plan my lessons.

(Lecturer 1)

Lecturer 1 teaches Radiation Physics by starting in the knower quadrant ('straightening the patient'), in contrast to Lecturer 2, who talks about radiation safety (in the élite quadrant) in relation to a depersonalized tumour.

Clinical educators' perspectives on first-year radiation physics

The clinical educators were not involved in the academic teaching of Radiation Physics but valued the role of the discipline in underpinning competent and safe practice:

You need to understand exactly why there is no room for error ... which is why radiation physics is so important. You can't just blindly push buttons you need to know ... why you're doing what you're doing.

(Clinical educator 1)

The clinical educators were aware that students had acquired a considerable knowledge of Radiation Physics. They described this as having 'head knowledge of radiation and what it entails' (Clinical educator 1). They were, however, sceptical of students' ability to apply the knowledge learned in the clinical context, as '...that comes with experience' (Clinical educator 1).

Revising the Threshold Concept Framework

Having studied participants' different understandings of first-year Radiation Physics from their various positions and experiences, elements of the Threshold Concept Framework were modified in the light of empirical data and insights provided by Specialization.

Explaining the (not entirely) boundedness of threshold concepts in Radiation Physics

Meyer and Land (2003, 2005) argue that it is the discipline-specific quality that makes threshold concepts difficult to learn and difficult to teach. The Radiation Physics lecturers were not in agreement about how 'bounded' the concepts of Radiation Physics were. Radiation Physics was recognized as a specialized sub-discipline of Physics, but it was also understood as an applied discipline developed for the treatment of patients. One of the lecturers described this bounded-yet-permeable nature of Radiation Physics as follows: 'I think it starts off in Physics … that's where the concept starts. It starts with concepts that are taught in Physics and so it does start there' (Lecturer 1).

In another version, its concepts are derived from practice, as another lecturer explained:

> And then we applied it … we went into the application straight away. In fact, what we did was we first went into that … there's two ways to look at radioactive decay. The description of it and then … the physics of it. The description actually we realized is independent of them having learned all this other Physics.
>
> (Lecturer 3)

This not-entirely-bounded nature of Radiation Physics characterizes many of its concepts. The students identified with the version of Radiation Physics that was closely tied to Radiation Therapy practice. A lecturer who taught a 'pure' version of Radiation Physics was said to be teaching 'Harvard University Physics' (Third-year student 4).

The notion of the bounded-yet-permeable is enhanced with the more precise descriptors of epistemic relations and social relations. Radiation Physics was not consistently described in terms of its epistemic relations to the discipline of Physics, and was, in fact, more often explained in terms of its application to Radiation Therapy practice. Radiation Therapy practice has stronger epistemic relations to Radiation Physics and stronger social relations to subjects. Patients are always at the centre of practice. Thus it makes sense for radiation therapists, who teach Radiation Physics, to understand Radiation Physics in terms of practice, rather than as a sub-discipline of Physics. This was evident in interviews with the clinical educators, most lecturers (especially lecturers who were radiation therapists), and among the students themselves, as is evident in the exchange between the interviewer and a senior student below:

INTERVIEWER: But if there's this one thing … what are [Radiation Physics concepts] … the must have?

SENIOR STUDENT: It has nothing to do with Physics, but I would say patient care is always number one (Third-year student 3).

Integrative (conceptual and practical)

Threshold concepts are said to reveal 'the previously hidden interrelatedness of things' (Meyer and Land 2005: 377). Threshold concepts build on prior concepts and once grasped enable the student to make connections between other concepts. This realization is often referred to as a 'light-bulb' or 'a-ha' moment (Cousin 2009). Lecturers in the study spoke about '…making connections to build knowledge' (Lecturer 3) and 'sequencing activities to build concepts' (Lecturer 3). These descriptions suggest that the threshold concepts in Radiation Physics integrate prior concepts learned in the discipline:

> It's impossible too for someone to understand [radiation physics], really understand it … without first understanding it conceptually. … If they don't have the conceptual understanding you never quite understand the Inverse Square Law, you never quite understand radioactive decay.
>
> (Lecturer 3)

But participants also proposed another version in which the concepts of Radiation Physics were integrated with practice. A first-year student explained her developing understanding in terms of integrating theory and practice:

> I think for me … it was the clinical part, like … going to the hospital and actually seeing it and experiencing what they are doing and I think that really brought it together.
>
> (First-year student 6)

The integrative nature of both non-threshold and threshold concepts in Radiation Physics refers to its 'hierarchical knowledge structure' (Bernstein 1999); that is, the concepts are cumulative; one concept is built on the other. It is difficult for students to acquire more advanced concepts if there are conceptual gaps in their understanding. As a Radiation Physics lecturer explained: 'I think it's got to do with conceptualization of basic principles that they are taught. Some people can't understand actually what we are doing' (Lecturer 3). Because Radiation Physics is so closely tied to Radiation Therapy, its integrative nature enables it to describe practice in particular ways. Radiation Physics is an applied discipline that describes the Physics of radiation treatment machines.

Temporarily troublesome

For first-year students, the concepts in Radiation Physics seemed, as Perkins put it, 'counter-intuitive, alien or seemingly incoherent' (Perkins 2006: 7). The students' troublesome experience was, however, temporary. Senior students, lecturers and clinical educators had mastered the once-troublesome concepts. Many could remember some of the difficulties that they had initially experienced. A senior student explained what encountering Radiation Physics for the first time felt like:

> I think if I look now at previous Physics lectures we've had … well, quite difficult, more difficult concepts that we haven't done in high school … so it's very difficult … [the lecturer is] talking about something there but you have nothing to reference it with. You have basically no idea what it's about really.
>
> (Third-year student 6)

Drawing on Specialization, the 'troublesome' nature of threshold concepts in Radiation Physics can be explained by a strengthening of the epistemic relations, described as a 'code drift' – that is, an occurrence in which a feature of a code is strengthened but not changed (ER↑) (Maton 2016: 237). Therefore, while Radiation Physics is characterized by a knowledge code (ER+, SR–), in threshold concepts epistemic relations are strengthened (ER↑). Understanding threshold concepts as rises in the epistemic relations enables us to separate the difficulty experienced by students as they enter the liminal zone, from the 'troublesome' nature of a discipline that has many threshold concepts, each of which represents an increase in the strength of epistemic relations. As an example of an increase in the strength of epistemic relations, Radiation Physics is used to develop algorithms for the three-dimensional geometrical plotting of the location of tumours and to plan the radiation dose to administer. The three-dimensional concepts that are embedded in Radiation Physics can cause students to experience difficulties:

> But [for] most students, it's just a difficult concept for them thinking, three dimensionally, where must a field come in? Just talking about maybe [organs at risk] a lung.... You know it's important to spare the other lung those kind of little stuff and that comes with experience ... you know ... where must a field go? How must it be labelled? Those simple type of things they struggle with.
>
> (Lecturer 1)

Radiation Physics is densely packed with non-threshold and threshold concepts which accounts for its being troublesome. It has ever-strengthening epistemic relations (ER+↑) comprising multiple non-threshold and threshold concepts, each of which needs to be mastered by the students before they can move on to the next one.

Liminality as encounters with radiation physics in theory and practice

Meyer *et al.* (2006) use the term 'liminality' in the sense of a 'rite of passage' that the student has to undergo before being accepted into a disciplinary community. Cousin (2006) describes how students often become 'stuck' and oscillate between understanding and misunderstanding. Most participants remembered their struggles with disciplinary concepts. In the excerpt below, a clinical educator recalls her struggles with Radiation Physics:

> Me personally, I panicked. I used to panic, you have to go read this, read that because the first question [the Supervisor is] going to ask you is how are you going to bring in your first beam? How are you going to place your first beam?
>
> (Clinical educator 7)

Land (2011) proposes that if the liminal space is not traversed, the student will only be able to perform in a 'ritualized manner.' This description is echoed in a clinical educator's account of the robot-like behaviour of some students, who seem to be stuck in this confusing space:

> ... there is something that I have picked up. The knowledge is there but the application of knowledge. ... For them, theory and practical, [are] two separate things. They know these things but to apply the knowledge in the clinical situation. It's like; it's a little bit far-fetched. As a result, what they do. ... I don't know which other words ... this might sound dramatic ... but it's like a robot issue. ... Because sometimes I ask a question, you do this, but why? Because ... you need to understand why am I doing it.
>
> (Clinical educator 2)

Land describes the liminal state as 'approximate to a kind of mimicry or lack of authenticity' (Land 2011: 176). This state is identified by Lecturer 2 who describes a student as going through the motions without comprehension:

> There is a missing link between the classroom and ... their technical environment, for sure. Because when you go to work then they stop thinking about the Physics. So, you just go and do your work, go and press the buttons, go and it's their day to day.
>
> (Lecturer 2)

Being in a state of 'liminality' is a characteristic of students learning threshold concepts. It should also be accepted that students will inevitably spend time in the liminal space in which they will experience difficulties in understanding, discussing and writing. The liminal space should be a safe space for students to learn from their mistakes (Land 2011). In LCT terms the liminality could be understood as recursive movements between weaker and stronger epistemic relations (ER↓↑).

Eventually irreversible

The idea of irreversibility was explained by the senior students as a gradual process of cumulative learning and gaining of insight: 'Radiation Physics ... then it just gets ... more clarity ... with every single time I got introduced to it again' (Second-year student 4). For many lecturers, for whom the concepts of Radiation Physics had long been internalized and irreversible, the idea of 'irreversibility' was evident in their frustration in trying to teach students something that was self-evident to them:

> I think the hardest thing to teach the students ... top of the list was x, y and z coordinates and understanding that x, y and z is not just one thing. So when I put the patient on the bed it's not just looking at midline and reference level and reference height. It's them translating that x, y and z to the x, y and z of the isocentre, which is a different x, y and z.

... It's a matter of explaining it and practising it and explaining and practising it and explaining it and practising it and then eventually a year down the line they'll understand it.

(Lecturer 1)

Threshold concepts are often described as 'irreversible,' but it is the student's attainment of the concept that is irreversible rather than the concept. Clouder (2005) for example, proposes that 'patient care' is a threshold concept in the health sciences and that 'the negotiation of a threshold is irreversible because experiences of caring are profound and are therefore not likely to be forgotten or unlearned' (Clouder 2005: 513).

Reconstitutive: The disciplinary underpinnings of practice

Meyer and Land's (2005) later inclusion of 'reconstitutive' as a threshold concept characteristic was an attempt to explain that when a student understood a threshold concept, there would be a shift in the student's 'mental models,' which is initially more likely to be noted by people other than the student, as Lecturer 1 explains: 'I think that is when the light-bulb moment comes, when you can amalgamate why you're doing that in planning and how you got the end result' (Lecturer 1).

Initially, Meyer and Land (2005) understood that it is students' thinking that is 'reconstituted' following the crossing of the threshold: 'What is being emphasized [in reconstitutiveness] is the inter-relatedness of the student's identity with thinking and language' (Meyer and Land 2005: 375). In a later work, however, Land *et al.* (2010) describe the threshold concept itself as 'reconstitutive.'

This reconfiguration occasions an ontological and an epistemic shift. The integration/reconfiguration and accompanying ontological/epistemic shift can be seen as reconstitutive features of the threshold concept.

(Land *et al.* 2010: iii)

Drawing on Specialization, a 'reconstitution' of the threshold concept would entail an understanding of its relationship to Radiation Therapy practice. Radiation Therapy has strong epistemic relations to Radiation Physics, as well as strong social relations to subjects. This suggests that disciplines can shift towards, or underpin practices. This is characteristic of applied disciplines in particular and was evident in much of the Radiation Physics lecturers' descriptions of their teaching, where they framed Radiation Physics concepts through the practice of Radiation Therapy:

What does it mean if I'm moving SUP? What does it mean if I'm moving INF? What is my x, y and z? How does the x, y and z apply to what my patient is doing or what I'm expecting the bed to do or...? and how that x, y and z, then relates to the treatment plan of the patient. ... So those bases help them understand not only the planning principles but also the set up principles which is the bread and butter of Radiation Therapy.

(Lecturer 1)

In LCT terms, applying abstract Radiation Physics concepts in practice involves a 'code shift' (Maton, 2016: 237) from weaker to stronger social relations (ER+,SR– → ER+,SR+).

Discursive: The specialist language of Radiation Physics

The 'discursive' dimension was also a later addition to the Threshold Concept Framework (Meyer and Land 2005). As threshold concepts would be likely to incorporate an enhanced and extended use of the language of the discipline and initially, lecturers found that:

Textbook terminology just goes straight over their heads I think sometimes. So I teach a concept the way I hope that they will understand and so in sort of layman's terms, I'll put up a presentation, showing them what I need for them to know with definitions in simple terms and we'll talk through it.

(Lecturer 1)

In time – and particularly with clinical experience – students started to use the disciplinary and professional discourse, as shown in the exchange between the interviewer and first-year student, who had returned from their first clinical rotation:

INTERVIEWER: Just -- what did you see and how did you do it?
FIRST-YEAR STUDENT 8: Oh, firstly you put the patient on the bed. Then you align the midline...
INTERVIEWER: What else after the midline?
FIRST-YEAR STUDENT 8: From the midline then you check the lateral tattoos. Then again, the midline.
INTERVIEWER: Can you see how you're starting to talk like them? Them ... the staff in the department and that's good. The more you do it the more confident you're going to become.

Meyer and Land (2005, 374) claim that the crossing of a threshold will incorporate an enhanced and extended use of language.

It is hard to imagine any shift in perspective that is not simultaneously accompanied by (or occasioned through) an extension of the student's use of language. Through this elaboration of discourse new thinking is brought into being, expressed, reflected upon and communicated.

Scientific discourses have developed within disciplines to represent complex disciplinary concepts, and these can be challenging for the newcomer, especially if the terms used also have everyday, non-specialist meanings. Cousin points out that mastery of a threshold concept can be inhibited by the prevalence of a 'common sense or intuitive understanding of it' (Cousin 2006: 5). Tan *et al.* (2019) warn that lecturers need to be careful with their use of 'anthropomorphic language' when discussing ionization energy and should consistently demonstrate the correct and technical language in their presentations and conversations with students (Tan *et al.* 2019).

The language of Radiation Physics requires stronger epistemic relations to the discipline and weaker social relations. First-year students find it difficult to remember the specialized terms and ways of communicating disciplinary knowledge, and would initially have weaker epistemic relations to Radiation Physics, but acquire the disciplinary discourse over time.

Transformative (knowledge and identity)

A threshold concept, once understood, causes a significant shift in the student's understanding, simultaneously with an identity shift. As Cousin puts it: 'New understandings are assimilated into our biography, becoming part of who we are, how we see, and how we feel' (Cousin 2010: 2). For the students in this study, these transformative shifts tended to happen in the clinical environment, rather than in the Physics classroom. A first-year student, recently back from her first clinical experience describes how the practice enhanced her conceptual understanding:

> And then by Linac 3, the referencing I understood better and even seeing it on the monitor and the calculations, you take the calculator and try to do it before. And then yes, that was what I have learned from there.
>
> (First-year student 10)

The clinical educators confirmed that transformative shifts were only likely to occur through practice:

> So ... say they're measuring a sep ... on the understanding that you ... measure from ant to post and ... they just don't get that – that's what they're doing. But the concept of what a sep is ... they know what it is.
>
> (Clinical educator 1)

In other words, students might know the concept of a sep (separation), but it is unlikely to become an internalized, irreversible or transformative concept until they have extended clinical experience. The clinical educators further cautioned that mastery of theoretical knowledge does not predict competent practice:

I think the type of student … because [they] are more confident … but they're not necessarily right. So, they are confident in the knowledge that they have with the studying. But then they think because they know that they automatically … can apply it … and they are very taken aback when they realize but they can't do it or they don't do it correctly.

(Clinical educator 2)

Reaching the point of transformative understanding through the integration of theory and practice is a long process:

And their time in the [clinical] department is different and their clinical exposure is different and I think what we want to see in a fourth year, we're possibly only going to see when they do community service.

(Clinical educator 6)

In the process of learning, the student changes, as Land *et al.* (2010) explain: 'the outcome of transformative learning … is that the content of the field of consciousness change' (Land *et al.* 2010: viii). Descriptions of the threshold concept as 'transformative' thus describe its effects, rather than its nature. However, in the same way that concepts can be 'reconstituted,' they can also be 'transformed,' such as in the 'code shift' (Maton 2016: 237) from Radiation Physics to Radiation Therapy, which was understood by a first-year student as the 'disappearance' of Radiation Physics in practice: 'Like when you work on the machines, you're not going to do any Physics there. It's just like in the background basically' (First-year student 11).

From the discussion above, we can locate elements of the Threshold Concept Framework on the Specialization plane (Figure 6.4). Radiation Physics is located in the 'knowledge' quadrant (ER+, SR–); threshold concepts in Radiation Physics are represented as a strengthening of the epistemic relations, or in LCT terms as a 'code drift' (Maton 2016: 237) (ER↑). Radiation Therapy is located in the élite quadrant (ER+, SR+) as it has epistemic relations to Radiation Physics and the necessary dispositions for clinical practice. Threshold concepts in Radiation Physics underpin practice, for example, the concepts of ionizing radiation underpin the practice of radiation protection in Radiation Therapy, the shift from the knowledge quadrant to the élite quadrant in LCT terms is a 'code shift' (Maton 2016: 237) from weaker social relations to stronger social relations to practice (ER+, SR– → ER+, SR+). Students' progress through the liminal zone is represented by the dotted line which moves from recursive learning to irreversible understanding of the threshold concept (ER↓↑, SR–); threshold concepts in Radiation Physics are represented as a strengthening of the epistemic relations, or in LCT terms as a 'code drift' (Maton 2016: 237) (ER↑). Radiation Therapy is located in the élite quadrant (ER+, SR+) as it has epistemic relations to Radiation Physics and the necessary dispositions for clinical practice. Threshold concepts in Radiation Physics underpin practice, for example, the concepts of ionizing radiation underpin the practice of radiation protection

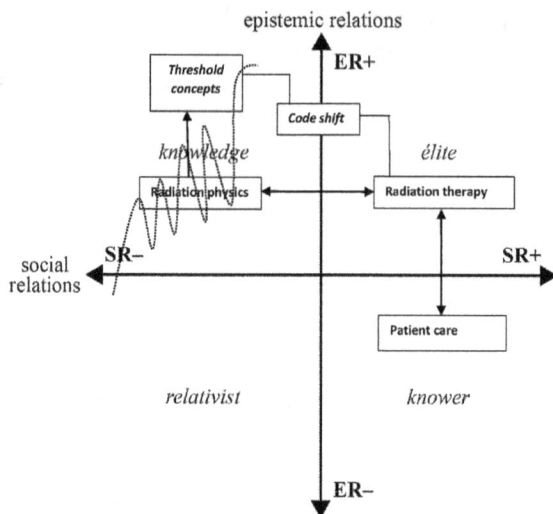

Figure 6.4 Plotting the Threshold Concept Framework on the specialization plane.

in Radiation Therapy, the shift from the knowledge quadrant to the élite quadrant in LCT terms is a 'code shift' (Maton 2016: 237) from weaker social relations to stronger social relations to practice (ER+, SR– → ER+, SR+). Students' progress through the liminal zone is represented by the dotted line which moves from recursive learning to irreversible understanding of the threshold concept (ER↓↑, SR–).

Detailed examples of the characteristics of threshold concepts with regard to their location on the Specialization plane are provided in Table 6.1.

Conclusion: An empirically grounded and theoretically consistent Threshold Concept Framework for Radiation Physics

This chapter set out to describe Radiation Physics in a theoretically consistent way for the purpose of benefitting lecturers, students and clinical educators. To address the research question, students, lecturers' and clinical educators' perceptions of Radiation Physics were elicited. These data were analyzed both with reference to the Threshold Concept Framework and Specialization. The engagement with empirical data and with theory enabled both a theoretically consistent and empirically grounded framework for the description of threshold concepts in Radiation Physics.

Through the analytical lens of Specialization, Radiation Physics was seen as having stronger epistemic relations and weaker social relations. The threshold concepts embedded in Radiation Physics were understood as increases in the strength of the epistemic relations, known in LCT terminology as an upward 'code drift' (Maton 2016: 237), thereby creating 'epistemological obstacles' (Meyer and Land 2005: 377) to student learning. In other words, those areas in which the epistemic relations become

Table 6.1 Using specialization codes to understand threshold concepts

Threshold Concept Descriptors	Using Specialization as Threshold Concept Descriptors	Codes	Example from the Data
Bounded	Radiation Physics is a Specialization of physics, the discipline is located in the knowledge quadrant.	ER+, SR–	*Physics is theory. It is what it is, what it is … you're teaching a concept (Lecturer 4).*
Integrative	Threshold concepts integrate prior concepts in Radiation Physics, represented as the strengthening of epistemic relations. Threshold concepts also underpin practice and need to be understood in terms of practice. This is represented as a code shift towards Radiation Therapy.	ER↑, SR– ER↑, SR+	*'if you don't actually understand the concept, you can't put a picture together of what is'* (Third-year student 3). *'…the linking of their book-based knowledge into clinical practice (Clinical educator 1).*
Troublesome	The epistemic relation strengthens in threshold concepts; this makes threshold concepts challenging or 'troublesome.'	ER↑	*When [clinical staff] mention SUP and moving from the reference to the isocentre…calculating that could be confusing at times (First-year student 6).*
Liminality	Liminality is explained as students' recursive attempts to understand the threshold concept.	ER↓↑, SR–	*I didn't understand a word he was saying because he's like very up there, clever with Physics and I'm like … don't understand (Third-year student 3).*
Irreversible	Students emerge from the liminal state when they grasp the threshold concept.	ER↑, SR–	*Radiation Physics … then it just gets … how can I put it … gets more clarity … with every single time I got introduced to it again (Second-year student 4).*

(*Continued*)

Table 6.1 (Continued)

Threshold Concept Descriptors	Using Specialization as Threshold Concept Descriptors	Codes	Example from the Data
Reconstitutive	Threshold concepts in Radiation Physics underpin Radiation Therapy practice, this is represented as a code shift on the Specialization plane.	ER↑, SR↓ ER↑, SR+	*when the light-bulb moment comes, when you can amalgamate why you're doing that in planning and how you got the end result (Clinical educator 2).*
Discursive	Discursive practices in an academic setting express strong epistemic relations, while in the clinical setting will have stronger social relations as well.	ER↑, SR±	*I think the difference between SSD and the different setups of the fixed Iso and the Iso on the patient itself (First-year student 5).*
Transformative	Transformation is understood as both understanding a threshold concept and being able to apply it in competent and safe practice. This could be understood as a code shift.	ER↑, SR↓ ER↑, SR+	*And then by Linac 3 ... the referencing I understood better and even seeing it on the monitor and the calculations, you take the calculator and try to do it before ... and then yes ... that was what I have learned (First-year student 10).*

stronger, cause students who are learning the discipline to experience them as 'troublesome.' Students then (usually temporarily) enter the liminal zone, where they experience confusion, but which is a process of recursive learning. As the students become more able to access and understand the strengthened epistemic relations of the threshold concept, they cross the threshold into clarity. When they venture into the clinical environment, they undertake a code shift (Maton 2016: 237) into the field of Radiation Therapy. In this shift, they have to move from an area of weaker social relations to one of stronger social relations, as they apply Radiation Physics in skilled and specialized practice. The students will also have to acquire the stronger social relations associated with patient care. All these aspects need to be taken into account by the Radiation Physics lecturers and clinical educators who will have to teach the difficult concepts in Radiation Physics.

References

Baillie, C., Bowden, J.A. and Meyer, J.H.F. (2013) 'Threshold capabilities: Threshold concepts and knowledge capability linked through variation theory,' *Higher Education*, 65(2): 227–246.

Barradell, S. (2013) 'The identification of threshold concepts: A review of theoretical complexities and methodological challenges,' *Higher Education*, 65(2): 265–276.

Baumann, M., Krause, M., Overgaard, J., Debus, J., Bentzen, S.M., Daartz, J., Richter, C., Zips, D. and Bortfeld, T. (2016) 'Radiation oncology in the era of precision medicine,' *Nature Reviews Cancer*, 16(4): 234–249.

Bernstein, B. (1999) 'Vertical and horizontal discourse: An essay,' *British Journal of Sociology of Education*, 20(2): 157–173.

Clouder, L. (2005) 'Caring as a 'threshold concept': Transforming students in higher education into health(care) professionals,' *Teaching in Higher Education*, 10(4): 505–517.

Cousin, G. (2006) 'An introduction to threshold concepts,' *Planet*, 17: 4–5.

Cousin, G. (2009) *Researching learning in higher education: An introduction to contemporary methods and approaches*, Routledge.

Cousin, G. (2010) 'Neither teacher-centred nor student-centred: Threshold concepts and research partnerships,' *Journal of Learning Development in Higher Education*, 2: 1–9.

Dunn, M.J. (2019) 'Crossing the threshold: When transition becomes troublesome for A-level students,' *Education and Health*, 37(1): 24–30.

Fredholm, A., Henningsohn, L., Savin-Baden, M. and Silén, C. (2019) 'The practice of thresholds: Autonomy in clinical education explored through variation theory and the threshold concepts framework,' *Teaching in Higher Education*, 25(3): 1–16.

Harnett, N., Bak, K., Lockhart, E., Ang, M., Zychia, L., Gutierrez, E. and Warde, P. (2018) 'The clinical specialist radiation therapist (CSRT): A case study exploring the effectiveness of a new advanced practice role in Canada,' *Journal of Medical Radiation Sciences*, 65(2): 86–96.

Harnett, N., Bak, K., Zychia, L. and Lockhart, E. (2014) 'A roadmap for change. Charting the course of the development of a new, advanced role for radiation therapists,' *Journal of Allied Health*, 43(2): 110–116.

Hudson, L., Engel-Hills, P. and Winberg, C. (2018) 'Threshold concepts in radiation physics underpinning professional practice in radiation therapy,' *International Journal of Practice-Based Learning in Health and Social Care*, 6(1): 53–63.

Kinchin, I.M., Cabot, L.B. and Hay, D.B. (2010) 'Visualizing expertise: Revealing the nature of a threshold concept in the development of an authentic pedagogy for clinical education,' in J. H. Meyer, R. Land and C. Baillie, (eds) *Threshold concepts and transformational learning*, Sense/Brill.

Körhasan, N.D. and Wang, L. (2016) 'Students' mental models of atomic spectra,' *Chemistry Education Research and Practice*, 17(4): 743–755.

Land, R. (2011) 'Crossing tribal boundaries: Interdisciplinarity as a threshold concept,' in P. Trowler, M. Saunders and V. Bamber (eds) *Tribes and territories in the 21st century: Rethinking the significance of disciplines in higher education*, Routledge.

Land, R., Meyer, J.H.F. and Baillie, C. (2010) 'Editors' preface,' in P. Trowler, M. Saunders and V. Bamber (eds) *Threshold concepts and transformational learning*, Sense/Brill.

Maton, K. (2014) *Knowledge and knowers: Towards a realist sociology of education,* Routledge.

Maton, K. (2016) 'Starting points: Resources and architectural glossary,' in K. Maton, A. Hood and S. Shay (eds) *Knowledge-building: Educational studies in Legitimation Code Theory,* Routledge.

Maton, K. and Chen, R. T-H. (2020) 'Specialization codes: Knowledge, knowers and student success,' in Martin, J. R., Maton, K. and Doran, Y. J. (eds) *Accessing Academic Discourse: Systemic functional linguistics and Legitimation Code Theory,* Routledge, 35–58.

Meyer, J.H.F. and Land, R. (2003) 'Threshold concepts and troublesome knowledge (1) – linkages to ways of thinking and practising,' in C. Rust (ed) *Improving student learning and practice – ten years on. Proceedings of the 2002 10th International Symposium: Improving Student Learning,* Oxford Centre for Staff and Learning Development.

Meyer, J.H.F. and Land, R. (2005) 'Threshold concepts and troublesome knowledge (2): Epistemological considerations and a conceptual framework for teaching and learning,' *Higher Education,* 49(3): 373–388.

Meyer, J.H.F., Land, R. and Davies, P. (2006) 'Implications of threshold concepts for course design and evaluation,' in J.H.F. Meyer and R. Land (eds) *Overcoming barriers to student understanding: Threshold concepts and troublesome knowledge,* Routledge.

Neve, H., Lloyd, H. and Collett, T. (2017) 'Understanding students' experiences of professionalism learning: A 'threshold' approach,' *Teaching in Higher Education,* 22(1): 92–108.

Nicola-Richmond, K., Pépin, G., Larkin, H. and Taylor, C. (2018) 'Threshold concepts in higher education: A synthesis of the literature relating to measurement of threshold crossing,' *Higher Education Research and Development,* 37(1): 101–114.

Park E.J. and Light G. (2009) 'Identifying atomic structure as a threshold concept: Student mental models and troublesomeness,' *International Journal of Science Education,* 31 (2): 233–258.

Perkins, D. (2006) 'Constructivism and troublesome knowledge,' in J.H.F. Meyer and R. Land (eds) *Overcoming barriers to student learning: Threshold concepts and troublesome knowledge,* Routledge.

Rowbottom, D.P. (2007) 'Demystifying threshold concepts,' *Journal of Philosophy of Education,* 41(2): 263–270.

Royeen, C.B., Jensen, G.M., Chapman, T.A. and Ciccone, T. (2010) 'Is interprofessionality a threshold concept for education and health care practice?' *Journal of Allied Health,* 39(Suppl 1): 251–252.

Scheja, M. and Pettersson, K. (2010) 'Transformation and contextualisation: Conceptualising students' conceptual understandings of threshold concepts in calculus,' *Higher Education,* 59(2): 221–241.

Tan, K.C.D., Taber, K.S., Liew, Y.Q. and Teo, K.L.A. (2019) 'A web-based ionisation energy diagnostic instrument: Exploiting the affordances of technology,' *Chemistry Education Research and Practice,* 20(2): 412–427.

Walker, G. (2013) 'A cognitive approach to threshold concepts,' *Higher Education,* 65(2): 247–263.

Wheelahan, L. (2015) 'Not just skills: What a focus on knowledge means for vocational education,' *Journal of Curriculum Studies,* 47(6): 750–762.

Weurlander, M., Scheja, M., Hult, H. and Wernerson, A. (2016) 'The struggle to understand: Exploring medical students' experiences of learning and understanding during a basic science course,' *Studies in Higher Education,* 41(3): 462–477.

Wolfson, A.J., Rowland, S.L., Lawrie, G.A. and Wright, A.H. (2014) 'Student conceptions about energy transformations: Progression from general chemistry to biochemistry,' *Chemistry Education Research and Practice,* 15(2): 168–183.

Zaky, H. (2018) 'Open your lesson gateways: Ways of thinking and teaching within the discipline,' *Journal of Modern Education Review,* 8(1): 109–117.

Part III
Biological Sciences

7 Interdisciplinarity requires careful stewardship of powerful knowledge

Gabi de Bie and Sioux McKenna

Knowledge boundaries

The world faces a number of problems that have unclear boundaries and which emerge from such a complex and shifting interplay of causes that these causes are almost impossible to fully identify, let alone address. These 'wicked problems,' as they are known, include the seemingly intractable issues of social injustice and environmental degradation. Universities are tasked with tackling such issues in two ways, through knowledge creation and through the education of young people. While universities are not the only social spaces reflecting on how best to address these issues, they are often the ones referred to in national policy as having this particular role to play.

There is, however, a concern that the ways in which universities organize themselves are not the best fit for undertaking this complex work (Mukuni and Price 2013; Van Duzer *et al.* 2020). There is a strong sense that the complexity of such 'wicked problems' requires an ability to work across the silos of traditional disciplines. Critics of the status quo often argue that the knowledge of the academy is stilted and segmented and that significant changes are needed for higher education to meet the demands of the era (Tully and Murgatroyd 2013; Thorne and Davig 1999).

In response to such calls, a number of curriculum innovations have been put in place to move from theoretical knowledge structured into traditional disciplines to the more concrete and interdisciplinary with a focus on the 'real world.' These educational innovations include problem-based learning, outcomes-based learning, competency-based learning and so on. Such innovations generally focus on what people will do with the knowledge they acquire. Teaching and learning is thus structured in ways that allow students to engage directly with how the knowledge of the academy plays out in the workplace. Instead of focusing on the abstracted principles of individual disciplines, students are given opportunities to engage in real-world cases and are expected to select theoretical knowledge to resolve the practical problem set in front of them.

These educational innovations entail integration of subjects that have traditionally been taught quite separately. Students are supported to work

DOI: 10.4324/9781003055549-10

across fields that have typically been inhabited by different researchers and different approaches to knowledge-making. The boundaries between disciplines classified into separate subjects are thus dismantled with the goal of developing learning that can work across artificial divides (Gerivani *et al.* 2020; Ghufron and Ermawati 2018).

Such innovative approaches have had many successes. Advocates of such innovations point to the greater levels of student engagement, the development of student autonomy and the extent to which students can enter the workplace and 'hit the road running' (Ge and Chua 2019; Ghufron and Ermawati 2018). There are indeed a number of benefits to approaches that more explicitly connect student learning to the practical implementation of knowledge and which allow students to move between disciplines that are traditionally carved into discrete 'subjects' on their timetables.

Critics of such approaches, however, raise a number of concerns, in particular, concerns about the more radical versions of such initiatives. The more extreme versions of these curriculum experiments seem to largely dismiss the idea that knowledge takes on different forms and that understanding how knowledge is made in different fields is key to the notion of 'powerful knowledge' (Shay 2013; Wheelahan 2007, 2009; Young and Muller 2013). These critics argue that too strong a focus on the immediately implementable can come at the cost of access to abstracted principles which allow us to move from a particular context to some future context, the likes of which we may not even be able to currently imagine. These critics argue that it is only if students have mastered the underpinning fundamentals of the disciplines that they can apply these across contexts in the workplace. The ideal curriculum, they argue, therefore sits somewhere in the middle: students are given access to the foundational knowledge and acquire an understanding of the abstracted principles underpinning such knowledge, and they are also exposed to a range of situations that require an application of such knowledge across the subject boundaries within which they may have studied it.

Hung (2019: 264) stresses that a 'critical element to successfully solve the problem [in a problem-based learning curriculum] is making sure that all disciplines have been taken into account,' but the literature on such educational initiatives provides little discussion on how the different forms of knowledge being integrated have been taken into account. Although Gerivani *et al.* (2020: 47) state that 'integration has been accepted as an important educational strategy in medical education,' Reddy and McKenna (2016) found, in their analysis of a problem-based learning medical curriculum, a lack of support for integration by academics. Klement *et al.* (2017) also note that a challenge to subject integration is ensuring support from faculty. While there are many possible reasons for resistance from academics who wish to hold on to traditional disciplinary divides, among them may be their sense of stewardship of knowledge, especially as there are very few deliberations in the literature about how the knowledge is structured within the constituent subjects of the curriculum.

It is important to note that while integration may be undertaken in the interests of interdisciplinarity – that is, to ensure better transfer of knowledge in the real world in which problems do not remain within disciplinary boundaries – such mergers are frequently undertaken for reasons of financial and logistical efficiency, which can ignore pedagogical implications. Klement *et al.* (2017), for example, indicate that the integration of Anatomy and other subjects in their study emerged at least in part as a requirement from their professional body and state that their goals for merger were to ensure better curriculum management and standardized examination, without any discussion on the nature of Anatomy and Physiology as sets of knowledge.

This chapter offers a case study of one such merger of disciplines: the merger of Anatomy and Physiology, into one subject, Human Biology, in the Faculty of Health Sciences at a South African university. Physiology and Anatomy are increasingly being taught together as one subject in various medical and allied health science curricula around the world (see, for example, Montayre and Sparks 2017), with some concerns being raised about whether students have sufficient time for all the constituent sub-sections (Rockarts *et al.* 2020). Many reasons are given for the integration of these two subjects, though in our literature search we found none that engage directly with the nature of the knowledge being integrated.

Case study of human biology

The larger study from which this chapter comes (De Bie 2016) tracked the curriculum of the two original subjects and the resultant merged subject from 1994 to 2013. The merged subject, Human Biology, was taught to students studying Occupational Therapy and Physiotherapy, who had previously been taught separately. Bringing together students studying towards different professions can allow shared learning between these related professions by students who would often work together in their future careers. Both the mergers of the subjects and the bringing together of the student body can thus be seen to have a clear and credible rationale. But, as this case study will show, where such mergers take place without due understanding of the different nature of knowledge in different disciplines, the results can be problematic and undermine the possibilities for cumulative learning.

This study asks the question: How does the structuring of the foundational Human Biology curriculum shape students' access to professional knowledge? The study explored whether the organization of the interdisciplinary curriculum of Human Biology served the fundamental needs of the two professions, and whether, as a matter of social justice, students' access to powerful knowledge was enabled by the form that the curriculum assumed. In order to interrogate the effects of merging two subjects, Anatomy and Physiology, into one, Human Biology, we drew on concepts offered to us by Legitimation Code Theory (LCT).

LCT concepts: Specialization and Semantics

The study drew on two LCT dimensions – Specialization and Semantics – in order to map out what was legitimated in the curricula of Anatomy and Physiology, and to then look at legitimation in the integrated Human Biology curriculum.

Specialization is used to identify the means by which a particular field of study legitimates knowledge and knowers in ways that are specific to it and differ from other fields (Maton 2014; Maton and Chen 2020). Specialization requires us to establish the extent to which the acquisition of specific forms of knowledge, practices and processes are central to the specialization of the field. This measure of the relations to the *object of study* is known as *epistemic relations*. Specialization simultaneously requires us to establish the extent to which particular dispositions or 'ways of acting, thinking or being' are required of knowers in order for them to be considered legitimate members of the field. This measure of the relations to the *subject of study* is known as *social relations*. Having established the nature of the epistemic relations and the social relations, we are able to map these onto a cartesian plane to establish the specialization code.

Given that this study looked at two subjects taught separately, Anatomy and Physiology, and then the curriculum after their merger to become one subject, Human Biology, Specialization allowed us to map the changes in the nature of what it is that was deemed to be legitimate. It should be born in mind that while the plane illustrated in Figure 7.1 highlights four principal codes – *élite, knowledge, knower* and *relativist codes* – there are an infinite number of positions within any of the quadrants. Furthermore, any area of

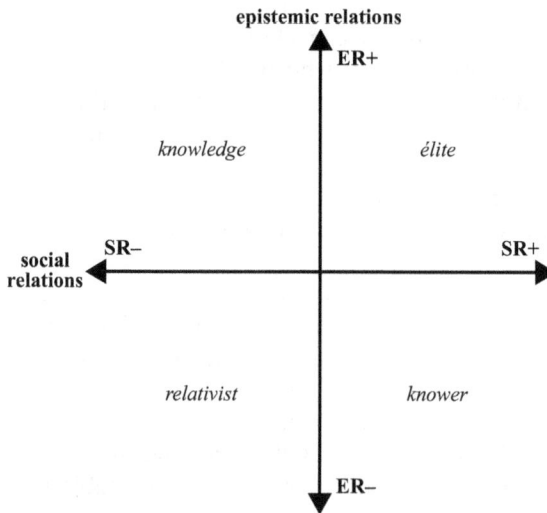

Figure 7.1 The specialization plane.

knowledge will undoubtedly inhabit various spaces, though it is likely that one will dominate. 'Code clashes' are where different positions come together in some way (Maton 2014); this can be where the understandings of the student and teacher differ regarding the expectations of a field of study, for example (Maton and Chen 2020). Specialization was used to interpret the epistemic relations and social relations, and thereby the specialization code, of the Anatomy curriculum, the Physiology curriculum and then the curriculum of the integrated Human Biology.

Semantics was also drawn upon in our analysis. Semantics is concerned with how the nature of meanings and offers two continua against which data can be mapped (Maton 2013, 2014, 2020). *Semantic density* conceptualizes the complexity of meanings condensed within knowledge practices. Here we shall use the concept to look at the ways in which meaning was communicated in curricula. Curricula which demand access to highly condensed terms and formulas are deemed to have stronger semantic density than those which rely more on everyday language. As a simple example, a cooking recipe might call for the addition of 'a cup of water,' and a Chemistry experiment in a school textbook might indicate '284mL H_2O'; the former has much weaker semantic density than the latter. The purpose of semantic density is not (or should not be) to make the text more difficult but rather to condense a lot of meaning into a text that can be readily communicated to other members of the field.

The other organizing principle in Semantics is *semantic gravity* (Maton 2009, 2013, 2014). This is an estimation of the extent to which the issue, concept or topic is tied to a particular context – that is, it has stronger semantic gravity – or whether the matter at hand can be applied across various contexts – that is, it has weaker semantic gravity. For example, the idea of semantic gravity can be considered as the extent to which what is being taught is connected directly to accessible real-world examples or students' own experiences (stronger semantic gravity) or whether what is being taught is more focused on principles rather than specific cases (weaker semantic gravity).

Ideally, teaching takes place in waves of semantic gravity where students are shown connections between (for example) more accessible real-world examples and more abstracted principled knowledge (line C in Figure 7.2). A flat-line of weaker semantic gravity (line A in Figure 7.2) can be problematic as students may battle to make sense of this highly abstracted knowledge if they cannot connect it to what they already know. A flat-line of stronger semantic gravity (line B in Figure 7.2) is equally problematic as students remain in the concrete realm of, for example, everyday experience or particular examples, without access to the powerful principles that would allow them to make sense of new contexts.

By using the tools offered by Specialization and Semantics, we were able to map the various ways in which legitimation was meted out in the Anatomy and Physiology curricula and the extent to which such legitimation shifted as the integrated Human Biology curriculum came into place.

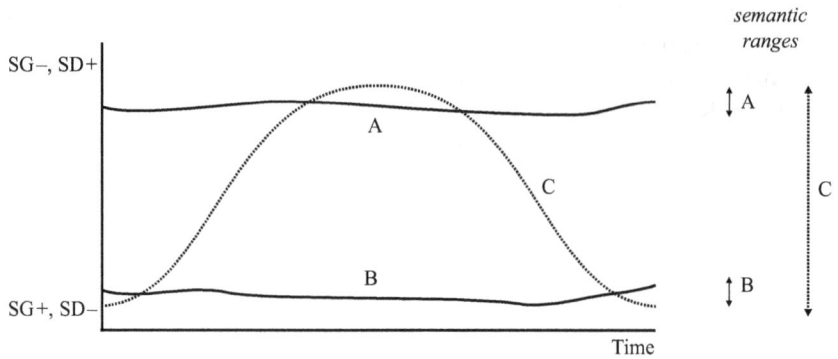

Figure 7.2 Three semantic profiles.

Research design

This case study is drawn from a much larger study that looked at data over a 20-year period, from 1994 to 2013, a period that included not only transitions in professional education but also extensive transformation in, and a different approach to, health delivery. In such a lengthy period, where both the fields of health and higher education saw enormous change, it is to be expected that this particular study would also evidence significant shifts. It is impossible to identify in the social world any particular macro, meso or micro shift that directly resulted in the curriculum shifts identified in this study. The realist position we take entails an understanding that events and experiences in the social world emerge from the complex interplay of multiple mechanisms (Archer 2000). This acknowledgement of epistemic relativism (Danermark *et al.* 2002) – that is, that our knowledge of the world is partial and subject to change on the basis of new information – should not be confused with ontological relativism, which suggests that all knowledge is personal and subjective. Using what Bhaskar (2016) refers to as judgemental rationality, as researchers we strove to identify the key causal mechanisms related to the effects on learning of the merger of Anatomy and Physiology to form Human Biology.

In particular, we were concerned with the function of knowledge itself. Knowledge is, somewhat ironically, often ignored in education research (Maton 2014, 2009). Common focus areas in educational research include the consideration of curriculum as a structure related to timetables and credits and so on, the consideration of students as individuals having a learning experience and the consideration of the university as a place of reproduction or disruption of social injustices. These are all important issues; however, there is a lack of focus on how knowledge differs from field to field and how such differences have effects on how the knowledge is taught and learned. This gap in much educational research has come to be known as 'knowledge blindness' (Maton *et al.* 2016).

Data in the form of Faculty Handbooks and Departmental Lecture Schedules for the 20 years under investigation were analyzed alongside detailed in-depth interviews with two lecturers in the Human Biology course, one who taught the Physiology sections and one who taught the Anatomy sections. Interviews were also conducted with a senior academic in Physiotherapy and a senior academic in Occupational Therapy. The data from these four interviewees is the focus of the case study presented in this chapter.

The two lecturers from the Human Biology course were interviewed for their understanding of what specializes the physiological and anatomical components of the Human Biology curriculum, what they considered as powerful knowledge for the professions and who they envisaged as the ideal knower. One Human Biology lecturer had Anatomy expertise and the other had expertise in Physiology.

The two lecturers from the two professional fields of Occupational Therapy and Physiotherapy were interviewed for their understanding of the extent to which the Human Biology curriculum prepared their students for each particular profession. While there are extensive overlaps between the professions of Occupational Therapy and Physiotherapy, they are distinguished in very specific ways.

Physiotherapy plays an essential role in helping people to maximize movement and to achieve optimal physical function. This involves consideration of the demands of daily living, and of occupational, recreational and sporting activities. Physiotherapists prescribe exercise programmes to promote physical activity and encourage an active lifestyle, which in turn contributes towards the prevention of health disorders. Physiotherapists are educated and trained to assess and treat a vast range of physical limitations and dysfunction by means of manual and electrotherapeutic techniques. In several countries of Western Europe, Australia and South Africa Physiotherapists are first-contact practitioners, which means that a referral from a medical doctor is not mandatory and a client can directly seek treatment from the therapist.

Occupational therapists believe that what people do every day has an important link with health and well-being. Illness or injury often disrupts people's ability to engage meaningfully in everyday occupations. Occupational therapists are trained to assess the person holistically, looking at all aspects of function, and analyze the environments where people live, work, play or pursue leisure activities so that they can understand how to improve function or adapt the environment in order to foster successful performance. Occupational therapy has developed various treatment modalities which enable people who have been ill, injured or disabled to recover their skills, or to develop new ones. In other words, occupation can be understood as being occupied in all facets of life rather than a concept of employment alone and the profession is centred on occupational functionality. Occupational therapists themselves admit to there being 'complex meanings and essential ties to human wellbeing ascribed to the concept of occupation' (Joubert 2010: 22) and are described by their Professional Board as working with anyone who has a permanent or temporary impairment in their physical or mental functioning and helping with rehabilitation of neuropsychological deficits including memory.

The lengthy interviews with the four academics provide the data from which this case study is developed. The interviews lasted over an hour, and in three cases, follow-up interviews were undertaken. They were semi-structured in that a short set of questions was sent to interviewees prior to the appointment and were used to guide the interview, but the process generally followed the form of a conversation with the interviewer asking probing questions and follow-up questions on the basis of what the interviewee raised. The interviews were recorded, with the interviewees' permission, and transcribed. Before they gave their informed consent, the interviewees were all fully informed of the research intentions, the data collection process and their rights to anonymity and to withdraw from the study.

This chapter considers the views of the four interviewees from the perspective of the structure of knowledge in Anatomy, Physiology and the Human Biology curriculum, and their views about how such knowledge is applied in the two professions of Occupational Therapy and Physiotherapy. The LCT tools described above were thus the analytical frames through which the data was considered.

Results

The data shows that the merger of Anatomy and Physiology to form the Human Biology course was undertaken to ensure coherence across these two core subjects. As one of the interviewees explains:

> The purpose of Human Biology is to provide our students with the core knowledge of Anatomy and Physiology that underpins everything that we then teach in the profession-specific courses later.... Because all of their profession-specific courses rely on that basic level of knowledge that we expect them to acquire in Human Biology.
>
> (Lecturer in Physiotherapy department)

There was evidence across the data that Physiotherapists and Occupational Therapists both need to draw on the knowledge of Anatomy and Physiology in integrated ways in their workplaces:

> So, on the Anatomy side they need to have a thorough knowledge of the Anatomy of the cardiorespiratory system. They also have to have a thorough knowledge of neuro-Anatomy to underpin the physiotherapy treatment assessment, application of techniques as applies to those systems. Physiology wise, they need to have a good understanding of cardiorespiratory Physiology you know … in their third year they do quite an intensive neurology course where they look at assessing and treating

head injuries, strokes (…) and they need to have a good grasp of Neurophysiology to understand the pathology on top of that.

(Lecturer in Physiotherapy department)

There was thus consensus that being able to draw on knowledge from both disciplines in an integrated way was key to the work of both professions. Despite this, the data revealed significant concerns about the extent to which the Human Biology course was a good fit for purpose. There was a concern that bringing these two disciplines together had led to gaps and schisms in students' learning:

> But my concern is that they're exposed to it at a level where we're expecting and basic underpinning knowledge that isn't there.
>
> (Lecturer in Physiotherapy department)

> But we would have a student who would not have understood the basics about joints, different kinds of joints.
>
> (Lecturer in Occupational Therapy department)

In this focus on the interviewee data, we offer two findings that we believe would be useful to take into consideration where similar curriculum changes are brought about in other programmes. The first finding was a concern about coherence within the newly merged programme and suggests that there were at times a code clash that was insufficiently considered in the curriculation of the Human Biology course. The second concern pertained to the extent to which the academics who offer the courses were consulted in the development of the merged curriculum. Each of these will now be discussed in turn.

Coherence and connection

There was agreement in the data that students preparing to work in the fields of Physiotherapy and Occupational Therapy would face problems in the workplace requiring an adept movement between knowledges:

> … I think the applications of a lot of the subjects obviously had to change with the current knowledge of, um, the current clinical knowledge that we have. Like the increase in diabetes you know, the advent of HIV/AIDS etc. So, so I think, I think the essential knowledge of the topics are important, have been tailored as we've gone through the last couple of years. I've tried to make it a bit more relevant in terms of the current clinical problems that we encounter now.
>
> (Physiology Lecturer in Human Biology)

Understanding that these problems cannot be addressed by drawing on the expertise of only one particular discipline is important for students. It was thus easy for the participants to acknowledge the justifications for bringing the two subjects together in ways that might allow students to see the complex interweaving of Anatomy and Physiology. Established disciplines with very clear boundaries can prevent students from making connections between them, and there is a need to ensure that the structures of the educational experiences do not prevent students from seeking creative understandings of and solutions to the intractable problems they face.

However, the dominant view emerging from the data was that the merger of the two subjects did not readily allow for such movement between the knowledges offered by each discipline.

> I've inherited a situation that was fragmented. I was told to only go to a certain level and then therefore the next year we continued. And even for myself, I found this very disjointed because when I spoke to the students and I said remember we covered this last year, all you sat with was 120 blank faces. And I found I had to reteach almost in essence the first part of the course to be able to continue with the second part of the course which for me is a waste of time. So, you not only have to remind them of old knowledge, you have to remind them of new knowledge. Because of the way we are forced to teach, because we have to do it section by section, students get the impression that blood vessels and nerves come section by section.
>
> (Anatomy lecturer in Human Biology)

There is a possibility that because the fields of production for both Anatomy and Physiology are very well established, there may be a resistance by academics inhabiting these fields to redraw their boundaries. Bernstein (2003) distinguishes between a *field of production* (where research is undertaken and knowledge is made), a *field of recontextualization* (where the curriculum is developed) and a *field of reproduction* (where teaching, learning and assessment take place). There are always conflicts within and between each of these fields, and the participants in this study did indeed understand the fields of reproduction of Anatomy to be very distinct from the field of reproduction of Physiology:

> [Learning Anatomy entails] ... to not only have to describe muscles, but try to integrate what is now going on around. Where this is sitting, in what region is it sitting, what defines that region. And then I also ask them to label diagrams. This is very important so that they know even though it's a 2D picture of what they're doing in 3D, it teaches them that everything has its own little region where it is going to be lying.
>
> (Anatomy Lecturer in Human Biology)

> I think they, they struggle with Physiology simply because of the nature of learning Physiology. Because it's much more conceptual ... it is not just an identification as for Anatomy. I think it's much more of an

applied, of an applied science to the concepts and I think that's why, and that's why they struggle because their learning methods that they come in with are not geared from, from day one. Their learning methods are not geared to learn in a, in a conceptual way.

(Physiology Lecturer in Human Biology)

The lecturers agreed that the nature of the knowledge in Anatomy and Physiology differed and that therefore the pedagogical approach differed too. This led to one lecturer suggesting that students of the merged Human Biology curriculum needed a guide to show how these two had been brought together:

A guide – a mind map – or a guidance to students on, so when we talk about movement it relies on this and that and you get this from, you know, Physiology lecturers and this from Anatomy lecturers and then we'll have a consolidation.

(Lecturer in Occupational Therapy department)

The extent to which the resistance to teaching the two subjects as one merged offering was from the desire to maintain separate territories cannot be established, but there was a strong sense expressed in the interviews that bringing the two subjects together restricted the flow of cumulative knowledge-building within each field of Anatomy and Physiology. A key rationale for the dismantling of traditional disciplinary boundaries is to ensure better coherence of knowledge so students should be able to draw on understandings from different fields of knowledge. In our data, we found the academics believed that the new subject made things even more fragmented because the students had not acquired underpinning principles of either discipline.

It is very confusing for students. I've heard complaints. But I said to them nothing is a stand-alone … because now they come with the stand-alone hand and all of a sudden, they have to learn and remember all of the other muscles that came beforehand.

(Anatomy Lecturer in Human Biology)

Clinical sciences is in crisis. It's not working at all and it concerns me greatly. For me the, the content that may still be missing. Because if students won't understand hypertension because the basics are missing, I'd have a problem. If students won't understand TB because the basics are missing, I'd have a problem.

(Lecturer in Occupational Therapy department)

At times, the structuring of the programme such that both Anatomy and Physiology lecturers focus on the same body area was complicated for purely pragmatic reasons:

... certain things to be taught at certain times but that policy cannot always be followed because I teach on other courses as well and therefore I'm only available at certain times so often sometimes we are out of sync. But we try to follow because they need the basis for Anatomy before they can do the applications in Physiotherapy or Occupational Therapy continuing of where the Anatomy has supposedly left off. But that in sync-ness doesn't always happen.

(Anatomy Lecturer in Human Biology)

By teaching the Anatomy and Physiology aspects of a particular part of the body in an integrated way in the Human Biology curriculum, it was hoped that students would be able to understand the full complexity of the human body. While both fields have a strong emphasis on objects of study and students are expected to engage with extensive knowledge (ER+), Anatomy has a very low emphasis on dispositions of the knower (SR–), whereas Physiology has a somewhat stronger emphasis on dispositions (SR–) as the student was expected to relate both as a propositional learner and as an applied, procedural scientist kind of knower. There are thus subtle differences in the ways in which the two fields are specialized.

Perhaps more problematically, greatly increasing the semantic gravity (SG+) by simultaneously looking at the intricacies of both the Anatomy and the Physiology of the hand, as per the example earlier, had the unintended consequence of decreasing the students' access to more abstract concepts (SG–) relevant across specific body parts or beyond particular ailments.

The muscles don't just start and end in a specific section. They cross the joints because obviously we know that as the muscle crosses a joint it moves that joint. So, it has implications for the other regions.... You cannot teach them piecemeal and expect the students to understand what is going on in those various areas. Like for example, your cardiovascular Anatomy is broken up by the respiratory Physiology sitting in the middle over there. It has to be taught – in a more integrated way because structure and function cannot be separated.... I look at the muscle, the origins and insertions. I look at what they do, how they work together as a group.... But you have to know what everything else is attached to and running through and what the support mechanism is in the body itself so that you know that everything works together. So that if there's a problem in the one area, it's going to have a knock-on effect for the rest of their systems going on around the skeleton.

(Anatomy Lecturer in Human Biology)

While the aim of disciplinary integration was repeatedly expressed in the data, the academics indicated that in practice the Human Biology curriculum was experienced as two discrete subjects.

So, the class test will have an Anatomy component and a Physiology component, Ja. But I'm saying within that paper, the ... even if it's Cardiovascular say and they did, you know, two weeks of Cardiovascular Anatomy and they did two weeks of Cardiovascular Physiology, there won't be one question that is an integrated question of both Anatomy and Physiology. It will be the Anatomy for 10 marks and Physiology for 10 marks.

(Anatomy Lecturer in Human Biology)

Let me put it plainly, the aim is integration but the final test is not integrated. There's no, there's no question about that. That's the honest side of it. And I think that it would need really the Anatomists and the Physiologists to really sit in a, in a real engaging type of way to come up with something.

(Physiology Lecturer in Human Biology)

Analysis of the curricula documents alongside the interviews allowed an understanding that the structure of the knowledge in the two disciplines, Anatomy and Physiology, differed in fairly significant ways (De Bie 2016). While both are hierarchical in the sense that new knowledge is typically added onto and subsumes prior understandings rather than transposing prior knowledge, they function in different ways. Physiology requires a deep understanding of systems in ways not required in Anatomy. This means that Anatomy can be taught in a more segmented way than Physiology. The Human Biology course, with its focus on a particular part of the human body thus worked fairly well for the more segmented knowledge of Anatomy but less so for the connected, system-focus of Physiology.

Because if you look at that, Anatomy has the overwhelming bulk of the lectures, you can understand that in a way because there's a lot of work to cover but remember Anatomy is structure, Physiology is function and structure and function must be fully integrated so they can understand the functionality of those various systems.

(Anatomy Lecturer in Human Biology)

I think they, they struggle with Physiology simply because of the nature of learning Physiology. Because it's much more conceptual, it is not just an identification as for Anatomy. I think it's much more of an applied, of an applied science to the concepts and I think that's why, and that's why they struggle because their learning methods that they come in with are not geared from, from day one. Their learning methods are not geared to learn in a conceptual way.

(Physiology Lecturer in Human Biology)

Following the merger, it was found that the ideal of disciplinary integration was not reached, and the segmental organization and structuring of the curriculum negatively impacted on cumulative knowledge-building. After the merger the disciplines to some extent lost their shape, and in particular, the hierarchical knowledge structure was compromised. By not having access to the necessary disciplinary knowledge structures and their associated practices, students' ability for scaffolding and integrating knowledge into the clinical arena was constrained.

> And for me one of the things in their final year is that it's underpinned because they're not secure in their basic knowledge. They don't trust their basic knowledge.
>
> (Lecturer in Physiotherapy department)

> ... students who may be too concerned about the structure and not so much about what the dysfunction of that body structure means in the big scheme of things.
>
> (Lecturer in Occupational Therapy department)

The organization of the current Human Biology curriculum was thus limited in its facilitation of cumulative learning. The merging of two subjects did not meet the goal of subject integration.

Curriculum input by academics

Curriculum decisions get made in the messy reality of society where any number of factors come into play. As indicated earlier, there are often international shifts and trends in education that are implemented with greater or lesser degrees of success across disciplines and geographical contexts. In the case of this merger, the data shows a sense that the decision-makers may not have ensured sufficient buy-in and understanding from those who became responsible for offering the course.

> As a component teacher on the Anatomy course, I do not make any decisions. I simply get told you are doing six weeks of ... and that's the basis of it. I also get given the book, so just see that everything within the book is covered... so there is in essence no guidance being given on the depth, the clarity and the amount of work that you put into it. I simply get told what to teach. I've not been invited to any curriculum decision meetings simply because I teach components of the courses.
>
> (Anatomy Lecturer in Human Biology)

This is a common problem in curriculum reform, where academics are expected to implement the decisions of others and may feel that they have not been appropriately consulted with the result that their understanding of the context, and their disciplinary expertise may be insufficiently considered. Reddy (2011) in her study on problem-based learning in a medical programme argued that without significant input in the field of recontextualization by those researching in the field of production and those responsible for teaching in the field of reproduction, curriculum experiments can easily be doomed.

In part this is because the academics teaching on the programme need to support the integrated course if they are to do it justice, and in part because academics might be able to point out distinctions in the knowledge structures being brought together and how these need to be taken into account.

> … but the scary part of it is that you could theoretically have a 40% Physiologist, 75% Anatomist and the student finishing on 65%. Now within that 65% if the Physiological knowledge going into the clinical, into the clinical third year is needed, you've got 40% Physiological basis that you're working with which is the, the scary part.
>
> (Physiology Lecturer in Human Biology)

> …if you lay the proper foundation, what you have to build on is that much steadier and that much all-encompassing than if you now suddenly have to start cramming in the third and the fourth year when they start going out and treating patients at the various clinics, they would either have that firm foundation they all have something good to build on. If that foundation is shaky and the Anatomy is shaky then it actually has a bad reflection on you in the coming years.
>
> (Anatomy lecturer in Human Biology)

The academics also expressed a concern about the teaching of future Physiotherapists and future Occupational Therapists the identical Human Biology curriculum in the same class. In keeping with discussions in the literature (French and Dowds 2008; Joubert 2010), both professions were identified in the larger study (De Bie 2016) as being having a very strong emphasis on the knowledge, skills and practices (ER+) at the same time as a strong emphasis on being a particular kind of knower, one who is compassionate and able to empathize with the patient (SR+); that is, they were both *élite codes* (Figure 7.3). There are, however, distinctions between them, with Physiotherapy having much stronger epistemic relations.

Furthermore, the semantic gravity is stronger in Occupational Therapy, which focuses on the patient's everyday context as the main concern, whereas Physiotherapy focuses on physical well-being more generally. There was a concern expressed that students would not understand the different ways in which that knowledge is drawn upon within their different, though connected fields:

> Physios are more clinical than OTs, they're more medical than OTs. OTs straddle the medical sciences and the social sciences. We don't have the tools to assess clinical conditions. It's not our focus. It's a different profession. Our basics aren't their basics.
>
> (Lecturer in Occupational Therapy department)

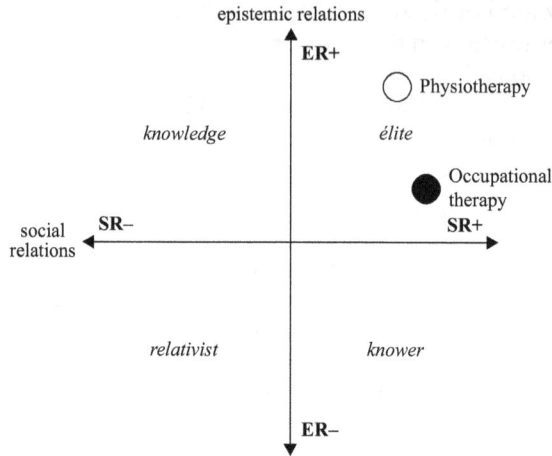

Figure 7.3 Specialization codes of the two professions.

There did seem to be an acknowledgement that there was 'increasing free-dom' (Physiology Lecturer in Human Biology) for the Human Biology lec-turers to make changes in the curriculum in the last few years and to work more closely with colleagues in the two target professions as they did so:

> So, the thing is, the, the question about the curriculum decisions and who makes them, is, is riveting because I think there has been a mind shift over the last five years in that.... And before that I think the bag-gage of the past was that, you know, Human Biology went on their own. They designed their course of Anatomy and Physiology and they ran it. Finished!
>
> (Physiology Lecturer in Human Biology)

It is therefore to be hoped that some of the concerns raised in this chapter can now be dealt with in this particular case. While few academics have the language with which to describe the structure of the knowledge and knowers legitimated in their courses, they may have a deep sense of what is needed in order to succeed in the field. Being able to articulate this, such as through the use of LCT, could be a strong starting point through which to engage in curriculum changes.

Conclusion

This case study offers the experiences of lecturers on a course that brought together two fields typically offered separately. It also considered the views of academics from the professional departments served by the merged course. All the academics were in favour of integration between subjects in ways that would allow the students, future health professionals, to draw from knowl-edge and practice across the separations of disciplines. However, it was clear that merging two fields into one course with one study guide, one timetable and one set of assessments is not a simple process. The nature of the expertise requires different lecturers and the content itself cannot readily be fused. LCT allows us to see that distinctions between the fields may be more than historical and relate to the structure of the knowers and knowledge. This needs to be taken seriously into account if discrete subjects are to be merged to form one offering.

The case study also considered the extent to which those offering the merged course, and those in the departments served by the course, were able to participate in the decision-making regarding the merger. Academics steeped in particular fields might be inclined to protect their territories and so make negotiations around curriculum structures difficult, but as experts in the target fields, they also have a strong understanding of the nature of their fields.

Unfortunately, few academics have a language by which to articulate the nature of legitimation in their fields, making it difficult for them to steward the powerful knowledge they have to offer. LCT offers a language by which academics can articulate what is valued and why this is so, and therefore possibly be more able to consider what should be changed and what should be retained and we prepare our students to take on the wicked problems of this complex world.

References

Archer, M. S. (2000) *Being human: The problem of agency*, Cambridge University Press.

Bernstein, B. (2003) *Class, codes and control: Theoretical studies towards a sociology of language*, Psychology Press.

Bhaskar, R. (2016) *Enlightened common sense: The philosophy of critical realism*, Routledge.

Danermark, B., Ekström, M., Jakobsen, L. and Karlsson, J. (2002) *Explaining society: Critical realism in the social sciences*, Routledge.

De Bie, G. (2016) 'Analysis of a foundational biomedical curriculum: Exploring cumulative knowledge-building in the rehabilitative health professions.' Unpublished PhD thesis. Rhodes University. https://core.ac.uk/download/pdf/145038435.pdf

French, H. and Dowds, J. (2008) 'An overview of continuing professional development in physiotherapy,' *Physiotherapy* 94(3): 190–197.

Ge, X. and Chua B.L. (2019) 'The role of self-directed learning in PBL: Implications for learners and scaffolding design,' in M. Moallem, W. Hung and N. Dabbagh (eds) *The Wiley handbook of problem-based learning*, Wiley Blackwell.

Gerivani, A., Sadeghi, T., Moonaghi, H.K. and Zendedel, A. (2020) 'Integrating of anatomy and physiology courses in basic medical sciences,' *Future of Medical Education Journal*, 10(4): 46–50.

Ghufron, A. and Ermawati S. (2018) 'The strengths and weaknesses of cooperative learning and problem-based learning in EFL writing class: Teachers and students' perspectives,' *International Journal of Instruction*, 11(4): 655–672.

Hung, W. (2019) 'Problem design in PBL,' in M. Moallem, W. Hung and N. Dabbagh (eds) *The Wiley Handbook of Problem-Based Learning*, Wiley Blackwell.

Joubert R. (2010) 'Exploring the history of occupational therapy's development in South Africa to reveal the flaws in our knowledge base,' *South African Journal of Occupational Therapy*, 40(3): 21–26.

Klement, B., Paulsen, D. and Wineski, L. (2017) 'Implementation and modification of an anatomy-based integrated curriculum,' *Anatomical Sciences Education*, 10: 262–275.

Maton, K. (2009) 'Cumulative and segmented learning: Exploring the role of curriculum structures in knowledge-building,' *British Journal of Sociology of Education*, 30(1): 43–57.

Maton, K. (2013) 'Making semantic waves: A key to cumulative knowledge-building,' *Linguistics and Education* 24(1): 18–22.

Maton, K. (2014) *Knowledge and knowers: Towards a realist sociology of education*, Routledge.

Maton, K. (2020) 'Semantic waves: Context, complexity and academic discourse,' in Martin, J. R., Maton, K. and Doran, Y. J. (eds) *Accessing Academic Discourse: Systemic functional linguistics and Legitimation Code Theory*, Routledge, 59–85.

Maton, K., Hood, S. and Shay. S. (2016) *Knowledge-building: Educational studies in Legitimation Code Theory*, Routledge.

Maton, K. and Chen R. (2020) 'Specialization codes: Knowledge, knowers and student success,' in J.R. Martin, K. Maton and Y.J. Doran (eds) *Accessing academic discourse: Systemic Functional Linguistics and Legitimation Code Theory*, Routledge.

Montayre, J. and Sparks T. (2017) 'Important yet unnecessary: Nursing students' perceptions of anatomy and physiology laboratory sessions,' *Teaching and Learning in Nursing*, 12(3): 216–219.

Mukuni, J. and Price, B. (2013) 'Portability of technical skills across occupations: A case for demolition of disciplinary silos?' *International Journal of Vocational and Technical Education*, 5(2): 21–28.

Reddy, S. (2011) 'Experiences of clinical practice in a problem-based learning medical curriculum and subsequent clinical environments,' Unpublished PhD thesis. University of KwaZulu-Natal.

Reddy, S and McKenna, S. (2016) 'The guinea pigs of a problem-based learning curriculum,' *Innovations in Education and Teaching International*, 53(1): 16–24.

Rockarts, J., Brewer-Deluce, D., Shali, A., Mohiadin, V. and Wainman. B. (2020) 'National survey on canadian undergraduate medical programs: The decline of the anatomical sciences in Canadian medical education,' *Anatomical Sciences Education*, 13(3): 381–389.

Shay, S. (2013) 'Conceptualizing curriculum differentiation in higher education: A sociology of knowledge point of view,' *British Journal of Sociology of Education*, 34(4): 563–582.

Thorne, D. and Davig, W. (1999) 'Toppling disciplinary silos: One suggestion for accounting and management,' *Journal of Education for Business*, 75(2): 99–103.

Tully, J. and Murgatroyd. S. (2013) *Rethinking post-secondary education – why universities and colleges need to change & what change could look like*, FutureTHINK Press.

VanDuzer, J. A., Leblond, P. and Gelb. S. (2020) 'Moving beyond disciplinary silos: Towards an integrated approach to international investment policy,' in J.A. vanDuzer and P. Leblond (eds) *Promoting and managing international investment*, Routledge.

Wheelahan, L. (2007) 'How competency-based training locks the working class out of powerful knowledge: A modified Bernsteinian analysis,' *British Journal of Sociology of Education*, 28(5): 637–651.

Wheelahan, L. (2009) 'The problem with CBT (and why constructivism makes things worse),' *Journal of Education and Work*, 22(3): 227–242.

Young, M., and Muller, J. (2013) 'On the powers of powerful knowledge,' *Review of Educational Research* 1(3): 229–250.

8 Advancing students' scientific discourse through collaborative pedagogy

Marnel Mouton, Ilse Rootman-le Grange, and Bernhardine Uys

Introduction

Scientists use the scientific method, which involves research questions and hypotheses: probable explanations based on observations. This is followed by meticulous design and execution of experiments and eventually the validation, refinement or rejection of the hypotheses (Carrol and Goodstein 2009). New findings are disseminated through publications where scientists argue the validity of their research among their peers, with the main aim of persuading their colleagues of the validity of their claims (National Research Council 2007). Their discourse presents logical arguments that aim to maximize the probability that readers will acknowledge the findings (Dyasi 2006). In general, scientific language displays objectivity by using abstract nouns derived from verbs and the third person passive voice, as well as numerous technical terms. It is, therefore, semantically dense and impersonal, and like all types of academic discourse, uses 'power words and grammar' to package the knowledge of the field (National Research Council 2007; Marshall and Case 2010; Martin 2013). This specialized language is, however, practically foreign to novices in the field (Marshall and Case 2010; Ambitious Science Teaching 2015).

Gee (2005) conceptualizes scientific language as one type of discourse, which he calls 'little d' discourse – the reading and writing typical of a certain community. This discourse may be very challenging for science students (novices). The second type is described as 'big D' Discourse, which presents the ways and values of a particular group or community, including reading and writing, but also 'behaving, interacting, valuing, thinking, believing and speaking' (Gee 1996; Marshall and Case 2010). Interestingly, the 'little d' discourse echoes the 'big D' Discourse thinking and valuing of the community. In the context of higher education, first-year science students are still newcomers, and Marshall and Case (2010) describe them as 'outsiders' to the language and practices ('little d' discourse and 'big D' Discourse) of the science disciplines. Their lecturers, on the other hand, are usually typically experienced scientists and therefore 'insiders' to the specific discipline, with its unique practices and academic d/Discourse (Marshall and Case 2010). Their role should thus be to induct their

DOI: 10.4324/9781003055549-11

students to become participants in the d/Discourse of the scientific community by promoting participation.

Higher education science classrooms are often more lecture orientated. Such practice may prevent participation, through argumentation, or even regular conversations about scientific topics. As 'insiders,' lecturers often regard the different ways to communicate the ideas in science (e.g. graphs, tables, representations) as self-evident, while the meanings of these may not be obvious to students because it is not made explicit in lectures and assessments. Marshall and Case (2010) therefore reason that 'in not allowing space for a critical engagement with these values and ways of thinking, numerous students are implicitly excluded from successful engagement with the subject.' Furthermore, in considering international practice, Case and co-workers (2013) argue for the importance of 'making explicit the academic literacy practices (d/Discourse) of the discipline' to advance learning for all students (here, academic literacy refers to the development of academic language skills and thinking strategies that are essential for successful study in the various disciplines). To achieve a more desirable outcome, therefore, educators need to make an effort to model scientific d/Discourse, and offer students opportunities to practice this specialized language. This includes actions such as scientific argumentation, using evidence to support knowledge claims, hypothesizing about scientific phenomena, writing up experiments and referring to data and patterns in data. When educators involve students in scientific writing, for example, the students engage in a metacognitive activity where they not only contemplate the correct wording to communicate their thinking but also reflect and clarify their thoughts in the process (Institute for Inquiry 2015). Such practice allows them to develop their discourse (scientific language) while fostering their scientific reasoning.

Various studies have looked at ways in which students' scientific argumentation and language can be developed. Engle and Conant (2002) proposed 'productive disciplinary engagement,' where connections are made between students' learning activities and the ways of scientific discourse (National Research Council 2007). Lee and Fradd (2002) argue for 'instructional congruence' where educators use students' language and cultural experiences to make science relatable, accessible and also meaningful (National Research Council 2007). This can be facilitated by providing students with opportunities to contemplate on and grasp new ways of thinking, with a balance between being challenged, yet feeling safe to experiment (ensuring that their norms and practices are valued). Other studies showed that the use of scientific language depends on students' everyday language established in their past or established over time (McNeill *et al.* 2005). Also, when science students must purposefully use language functions to articulate science, their content knowledge, as well as their language and mathematical proficiency, have been shown to improve (Dyasi 2006). This is due to its role as a key cognitive tool in the development of problem-solving and higher-order thinking. Kelly-Laubscher and co-workers (2014, 2017) highlighted the importance of exposing students to examples of the level of scientific writing they are expected to produce and assisting them in deconstructing these texts to better understand what is expected of them. Thus, educators need to draw on the present strengths of students, raise their awareness of the various types of discourse and make connections between them, and also make explicit what is expected from them. This way, science students will learn the 'rules of the game' and that scientific discourse is distinct for the purpose of building theories, interpreting data and logically communicating new findings. And also, that explanations and claims in science always need to be supported by rigorous scientific evidence (McNeill *et al.* 2005).

Scientific writing is demanding for the majority of students, including English-speaking students writing in their home language (McNeill *et al.* 2005). Chimbganda (2000) showed the various strategies that first-year Biology students, with English as a second language, use to compensate for their limited writing proficiency. And Clarke (2015) reported on the influence of first-year students' prior writing experiences at the school level on their word and grammar choices. Moreover, it was found that students require more skills to communicate their scientific thinking in written form than in verbal form due to a higher level of language skill needed for writing than for speaking (McNeill *et al.* 2005; Institute for Inquiry 2015). This implies that students' language proficiency will have an impact on the development of their scientific discourse. In South African schools, the language of instruction (mostly English), is often different to the learners' spoken home language, which further impacts the development of their scientific discourse. According to Boughey (2002), problems arise because students struggle to 'manipulate the forms of the [language of instruction] in a way that would allow them to receive and pass on the thoughts developed in the disciplines.' Some authors consequently argue for a pedagogy that will recognize that students oftentimes may not have the necessary language skills required to succeed in some disciplines, such as the sciences (Hurst 2010; Kirby 2010). Moreover, Maton (2013) showed that there is often a disconnection between complex disciplinary reading or 'high-stakes reading,' and the production of appropriate discourse or 'high-stakes writing.' To address some of these issues, many institutions of higher education have implemented independent (add-on) academic literacy courses and modules (Boughey 2002; Jacobs 2007). However, a host of studies have shown that academic literacy is taught more effectively when combined with disciplinary content or 'literacy across the curriculum,' compared to the non-integrated approaches (Boughey 2002; Jacobs 2007; Case *et al.* 2013). Jacobs (2007) argues that disciplinary discourse should be made explicit to students by their lecturers, while simultaneously introducing them to the forms of inquiry and knowledge production of the specific discipline. Kirk (2019) corroborates this by stating that academic literacies curricula are not neutral for the communication of academic knowledge and that language studies should not be presented separately from discipline content to advance learning for all students.

In this study, we focus on first-year Biology students' scientific discourse skills, and we explore ways to develop this fundamental skill. Here, discourse refers mostly to 'little d' discourse but also implicitly to 'big D' Discourse. Key goals were to help students' bridge the gap between reading complex discipline content and writing scientifically ('little d' discourse), and equally important, to make the ways of inquiry and knowledge production in Biology explicit to the students ('big D' Discourse). To develop students' discourse skills, we followed the scholarly approach of 'collaborative pedagogy' (Jacobs 2007) by cooperating with our colleagues, the academic literacy lecturers. The idea was for these lecturers to help our students gain mastery over textual choices for, in this case, Biology knowledge practices. Together, we designed a collaborative project to provide the first-year Biology students with an opportunity to develop their scientific language skills, through an introduction to the forms of inquiry and knowledge production of the specific discipline of Biology. Thereafter, we evaluated their use of scientific discourse from the following summative assessment. Also, we investigated the level of scientific discourse found in their prescribed first-year Biology textbook compared to the level of discourse found in the current high school textbook.

Theoretical framework: Semantics dimension of Legitimation Code Theory

To formulate and design the project as well as analyze the data, we drew on the Legitimation Code Theory (LCT) concept of *semantic density*, which explores the degree of complexity of meaning (Maton 2013, 2014a, 2014b). LCT is a realist framework that considers knowledge practices. It is a multidimensional toolkit that offers different dimensions to analyze particular sets of organizing principles which underlie practices. LCT conceptualizes the complexity of meaning as *semantic density*, which can be weaker or stronger along a continuum and can also be weakened and strengthened in practice.

Scientific language is generally complex and therefore represents stronger semantic density. However, complexity is a relative term, and often simply refers to the cognitive demand of the assignment. In contrast, semantic density affords greater specificity, conceptualizing complexity in terms of the condensation of meaning within practices, where condensation refers to adding meaning to a term or practice. In this chapter, we worked with *epistemic–semantic density* (ESD), which deals with epistemological condensation of formal disciplinary definitions and descriptions (Maton and Doran 2017). Epistemic–semantic density further explores the relationality of meanings. Thus, the greater the number of relations to other meanings of terms or concepts, referred to as a *constellation* of meanings, the stronger the epistemic–semantic density (Maton 2013; Maton and Doran 2017).

Several studies have investigated teaching and learning in Biology using Semantics. Kelly-Laubscher and Luckett (2016) showed clear differences in curriculum structure between high school and university Biology. Others showed how this gap can be managed by using LCT tools to plan and execute various interventions (Mouton and Archer 2018; Mouton 2019). Other studies have shown that shifts between more complex and simpler meanings (stronger and weaker epistemic–semantic density) are crucial to support cumulative knowledge-building (Maton 2013, 2014a, 2014b). Martin (2013) further showed that complex language choices are associated with these semantic shifts.

In practice, we expect students to express their subject knowledge using discourse ('little d' discourse). However, they often struggle to formulate their responses scientifically, using both simpler and complex meanings. Therefore, the rationale of this chapter is to explore ways of teaching students the 'rules of the game' in scientific writing, using the concept of semantic density, before their summative assessments in which appropriate scientific discourse is expected. In classroom activities, we often use terms such as 'power language' to communicate to students how to formulate scientific discourse. Thus, this study aimed to help students build their Biology knowledge and power language by participating in classroom activities that would facilitate the use of strengthening and weakening semantic density. This approach may assist students to connect the two types of language, mundane and scientific talk, through formulation practice and recontextualization (Skovholt 2016).

Methodology

In this chapter, we explored ways to develop the scientific discourse of first-year Biology students, for both their reading and writing, and thereafter we studied their use of this fundamental skill in a summative assessment. The Biology lecturer, therefore, identified a section in the first-year curriculum where students have struggled with articulating their conceptual understanding using appropriate scientific discourse, especially during written assessments. To develop the students' scientific writing skills, a project was designed using collaborative pedagogy, thus cooperation between the disciplinary lecturer (Biology) and the academic literacies lecturers, who also teach the same group of students' general scientific communication skills in a separate module. During the project, students worked in groups of three, researching specific structures found in eukaryotic cells. To practice scientific writing, they had to compile a written report about the structure and function of the organelles and systems found in these cells. The next step was to submit the report to the academic literacies' lecturers for feedback on the language and grammar of their writing. After revising their reports, they then submitted the final version to the Biology lecturer who assessed the scientific content, argumentation and discourse.

To analyze individual students' use of scientific discourse after the developmental opportunity, the final written assessment of the semester was used. Six students were selected, representative of the cohort, considering their achievements, their Biology background, as well as their language proficiency. The summative assessment covered most of the semester's content and also included the content of the project and written report. The discourse that we analyzed for this chapter focused on the nucleus and nuclear envelope, a section of the prescribed first-year Biology curriculum. To further contextualize our study, we also analyzed the corresponding content from the school and first-year Biology textbooks.

Participants

Students enrolled in the Extended Degree Programme (EDP) Biology module (Biology 146; instructed in English) at a South African University, participated in the project. The EDP had been implemented for students from previously disadvantaged backgrounds who fall just short of the university's programme entry requirements for mainstream offerings. The summative assessments of six students, representing various levels of academic proficiency were selected for the semantic analysis. An aspect that has to be taken into account is the fact that not all of the students in this group took Biology as a school subject. Moreover, a significant number of these students received secondary education in languages other than English. English, which is the language of instruction in this module, was often their second language.

Developing a translation device for semantic density analysis

Maton and Doran (2017) proposed a 'generic translation device' for analyzing how epistemic–semantic density (ESD) realizes in English discourse. They offer different tools for individual words, word-grouping, clausing and sequencing. Here we have related the translation device for wording to scientific discourse as Figure 8.1. This device can be used to analyze the complexity of meaning expressed by words in the discourse and how meaning is added or increased through combining words with additional words. For this study, we used both the wording and word-grouping tools (Biology examples have been included in Table 8.1).

Wording tool

The wording tool is divided into two broad categories or types – namely, *technical* words and *everyday* words. The meanings of technical words are often 'assumed within their specialized domain unless located in pedagogic settings' where their technical character may have to be emphasized. Moreover, technical words carry significance to specialists in the field, whereas they may appear foreign or dense to non-specialist readers. They are often nouns, longer words, names or place names with clearly defined meanings and strictly defined relations to specific contexts (Maton and Doran 2017). As a result, technical words are placed at the more complex, stronger end of the epistemic–semantic density continuum (ESD+).

Table 8.1 Epistemic–semantic density categories of Maton and Doran (2017) with descriptions and examples from students' discourse

	ESD category	Sub-subtype category	Description and examples from student discourse
ESD+ ↑	Technical; Conglomerate; Properties	8	This group typically includes actions and processes with multiple distinct parts each with its technical meaning. E.g. 'assisted exchange' and 'protein synthesis.'
	Technical; Conglomerate; Elements	7	This group also contains concepts with multiple distinct parts but does not include processes or actions. E.g. 'nuclear envelope' and 'eukaryotic chromosome.'
	Technical; Compact; Properties	6	This group typically includes actions and processes but with a single meaning. E.g. 'shuttling' (cargo) and 'expression' (gene).
	Technical; Compact; Elements	5	This group includes concepts with a single technical meaning. E.g. 'nucleus' and 'membrane.'
	Everyday; Consolidated; Specialist	4	This group contains concepts that are used in everyday language but in this context is dominated by specific technical meaning. E.g. 'hereditary information' and 'genetic material.'
	Everyday; Consolidated; Generalist	3	This group contains concepts that are used in everyday language but in this case has a more general technical meaning. E.g. 'signal' (noun) and 'molecule.'
	Everyday; Common; Nuanced	2	This group includes concepts that are used in everyday language and represent single happenings or qualities. E.g. 'reinforced' and 'embedded.'
ESD−	Everyday; Common; Plain	1	This group contains concepts that are used in everyday language with relatively general meaning. E.g. 'separate' and 'line.'

On the other end, *everyday* words represent simpler meanings and therefore weaker epistemic–semantic density (ESD−). The meaning of everyday words is not set in specialized fields and they are generally judged based on their usage in more common contexts. They are often shorter words in comparison to technical words and can be any word type (nouns, verbs,

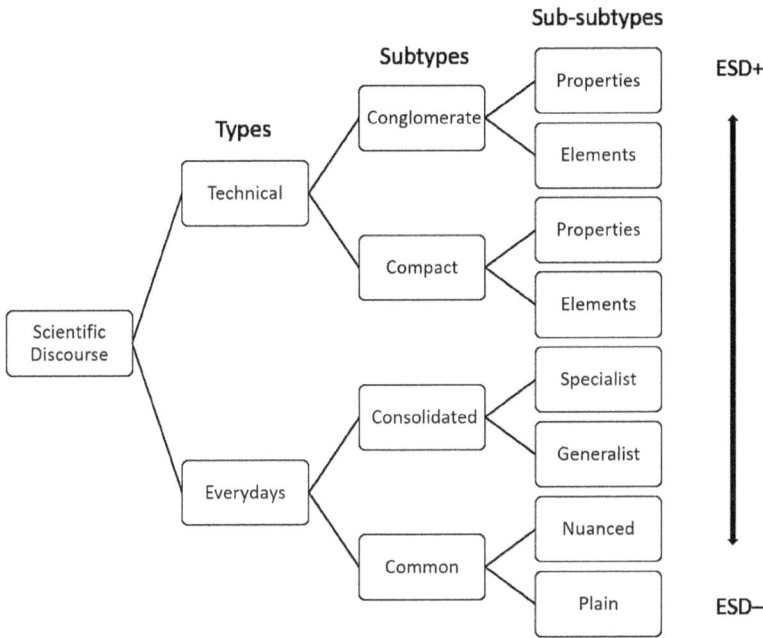

Figure 8.1 Wording tool for epistemic–semantic density in English discourse (Maton and Doran 2017).

adjectives, etc.). Everyday words can be related to a wide range of words without losing meaning, creating more fluid relations to various contexts.

At the next level of the wording tool, the two types are further subdivided to create four subtypes, which are then further subdivided to end up with eight sub-sub-types, as shown in Figure 8.1.

Subtypes of technical words

Technical words comprise *conglomerate* words and *compact* words. Conglomerates, as the term suggests, are words containing more than one part or concept, each one having a technical meaning. In contrast, compacts are single units with a single technical meaning. Conglomerates are there-fore considered to be more complex (have stronger ESD) compared to compacts, as they contain more meaningful parts. The sub-sub types of both conglomerates and compacts provide an even finer level of ESD analysis. Elements refer to 'an item, entity or thing of some kind,' whereas properties refer to 'an action or quality of an item, entity or thing.' Conglomerate properties are therefore seen to be more complex (stronger ESD) than con-glomerate elements.

Subtypes of everyday words

The sub-categories for everyday words are *consolidated* words and *common* words. Consolidateds encode 'happenings or qualities as things' (e.g. invade and invasion) while commons remain as they are, 'happenings' or 'qualities.' The term 'happening' refers to processes or events normally presented by verbs. Similarly, 'things' refer to items or elements represented by nouns. The finer level introduced for consolidateds distinguishes between *specialist* words and *generalist* words. Specialist words are *consolidateds* set in text dominated by technical words. In contrast, generalist words are *consolidateds* found in text dominated by everyday words. Two subtypes of common words can also be distinguished: *nuanced* and *plain* words. The former refers to words that exhibit more differentiated meanings whereas the latter are relatively general and simpler in meaning, e.g. 'embedded' (nuanced) vs 'lying in' (plain).

Proxy words

These are stand-in or replacement words used in text, such as 'nucleus' and 'it,' where 'it' refers to the nucleus in this case, or the term that was used earlier in the text. In terms of complexity (ESD), proxies are not as strong as the original word but not much weaker either. This is due to proxies not exhibiting the same perception of complexity, even though they represent the original word.

Word-grouping tool

Where the wording tool of Maton and Doran (2017) allows one to rate individual words in terms of epistemic–semantic density, the *word-grouping* tool considers the effect of combining or grouping words on ESD (Table 8.2). When words are grouped, they can often strengthen ESD. Maton and Doran's (2017) word-grouping tool describes three types of word groupings, called modifications, that increase the strength of a word's ESD:

Table 8.2 An adjusted version of the word-grouping tool for epistemic–semantic density (Maton and Doran 2017)

Type	ESD as proposed by Maton and Doran (2017)	How it was used in this chapter; our translation device
located	ESD↑	ESD↑
categorized	ESD↑↑	ESD↑
embedded	ESD↑↑↑	ESD↑
defined	not included	ESD↑

- *Located* modifications (ESD↑) increase meaning by specifying a specific location in time or space, e.g. 'structures form <u>around the genes</u>.' This allows for further differentiation.
- *Categorized* modifications (ESD↑↑) increase meaning by specifying a distinct type of word, e.g. '<u>Free-floating</u> nucleotides,' '<u>unwound</u> DNA.' This allows for further differentiation by indicating the specific type and subtype.
- *Embedded* modifications (ESD↑↑↑) increase meaning by showing that the word is active in an event or process, e.g. 'genes <u>that are responsible for storing</u> the hereditary information.' Thus, this type of modification specifies a specific type of secretion plus a specific type of activity.

Modifications can also be combined. The more modifications that are added in the text for a specific word, the stronger the ESD.

Results and discussion

Biology, like all other academic discourse, uses 'power words and power grammar' to elucidate the knowledge of the field. Martin (2013) describes 'power words' as technical terms with a 'greater strength of semantic density' and 'power grammar' as the 'knowledge construing power of grammatical metaphor' (which is one way of strengthening ESD). Thus, a discipline's knowledge gets packaged into text that stores the descriptions of the knowledge field. Each discipline is characterized by a unique genre, and students need to master the unique power composition of each discipline to know how to scaffold and organize these genres for the particular discourse, especially in written assessments. According to Martin (2013), power composition incorporates both power words and power grammar to organize writing that regularly shifts between complex meaning (ESD+) and simpler meaning (ESD−). This leads to academic writing that is precise, critical and objective, composed of complex meaning that is based on concrete evidence.

The textbooks

We examined the discourse of the two textbooks familiar to our students: the prescribed first-year textbook used in this module and the prescribed school textbook from their previous learning. Students must be able to access the knowledge in these sources by reading, and these textbooks would also serve as models for students to scaffold and organize the genres for their Biology discourse. We reasoned that understanding the discourse profiles of these texts in terms of power composition would serve as a baseline for comparing the students' discourse. Our analyses showed that the scientific discourse used in these two textbooks differed significantly with regard to complexity, density and volume (Figure 8.2). In Figure 8.2, complexity of meaning is weakest at the bottom of each bar and becomes stronger towards the top (Sub-subtype categories 1 to 8). Thus, the bottom band is weakest and the band second from the top representing the most complex meaning. The

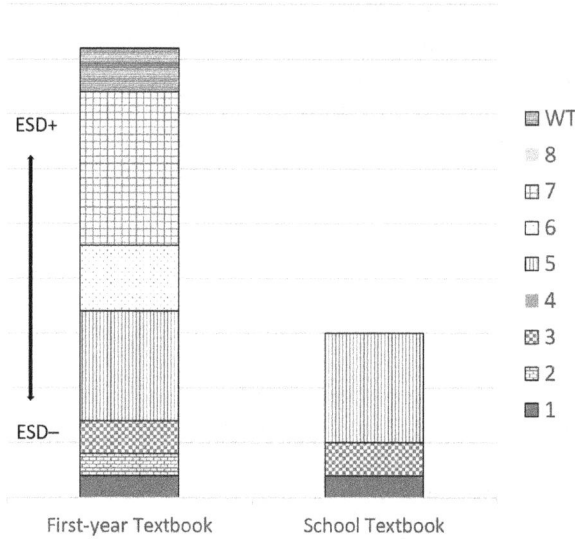

Figure 8.2 Bar graph indicating proportion of simpler to more complex meaning in the respective descriptions of the nucleus of a eukaryotic cell from the first-year and the school textbooks.

band on the top of each bar represents terms that strengthened the complexity of the descriptions (word-grouping tool). Although a difference between these two resources seems obvious and was therefore anticipated, we completely underestimated the magnitude of the variance, both quantitative and qualitatively. Firstly, the volume of text in the first-year textbook is significantly greater than that of the school textbook. Moreover, the first-year textbook uses a wide range of words and terms, with many coming from the two high-end, more complex meaning categories 6 and 7 (Technical; Conglomerate; Elements and Technical; Compact; Properties). Terms from these two categories (6 and 7) are completely absent from the school textbook. Thus, the first-year textbook uses far more words (quantitative) and significantly more power words and compositions (stronger ESD; qualitatively) than the school textbook, which is very elementary in comparison. Moreover, there is significant epistemological condensation from the school to the first-year textbook and curricula. *Condensation* refers to the process of adding meaning, in this case to biological terms (Maton and Doran 2017), where substantial meaning is added to terms they have learnt in school. Thus, when first-year students engage with their new curriculum and textbook, they are suddenly confronted with an exceptionally steep increase in volume, as well as complexity and density in meaning, which includes an increase in the number of power words and added meaning to known terms. Many newcomers are underprepared for this steep learning curve. Valencia (2014) found a similar situation with high school learners struggling with textbook reading and argues that these texts are 'not structured like any other authentic reading.' The students who read these textbooks already have to deal with learning many new concepts and data. Moreover, they are also confronted with new ways of thinking and reasoning that are important to the subject matter. Many of the students have not developed the necessary comprehension skills for learning from such text. Valencia (2014) therefore argues that 'not only do [the students] need to learn the content, they also need to learn how to learn from complex subject-matter texts.' Our study revealed a similar situation in this first-year cohort.

Students' scientific vocabulary

We found that the scientific vocabulary and use of power words varied considerably among the students in the summative assessment. Figure 8.3 shows that three of the six students (Students 4 to 6) displayed a proficient command of the Biology vocabulary (power words with stronger ESD) and grammar needed to describe the structure and functions of the nucleus of the eukaryotic cell. The remaining three students (Students 1 to 3) struggled to effectively portray this biological structure and all its components using written discourse. In Figure 8.3, complexity of meaning is weakest at the bottom of each bar and becomes stronger towards the top (Sub-subtype categories 1 to 8). Thus, the bottom band is weakest and the band second from the top representing the most complex meaning. The band on the top of each bar represents terms that strengthened the complexity of the descriptions (word-grouping tool). When comparing the students' descriptive accounts to the first-year textbook, it was encouraging to see that Student 4 managed to use most of the terms from the textbook, and further demonstrated a sound understanding of how these components fit together and relate to one another (power composition), as shown by the concept map in Figure 8.4a (a tick

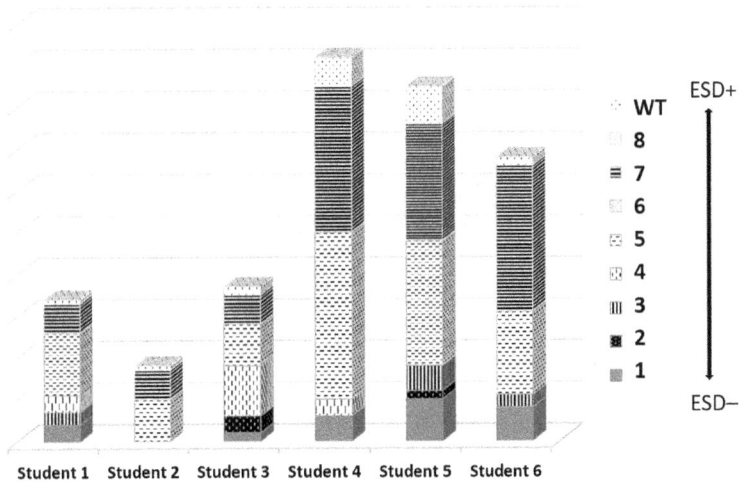

Figure 8.3 Bar graph indicating proportion of simpler to more complex meaning in the respective descriptions of the nucleus of a eukaryotic cell from the students' (1 to 6) final summative assessments.

160 *Marnel Mouton et al.*

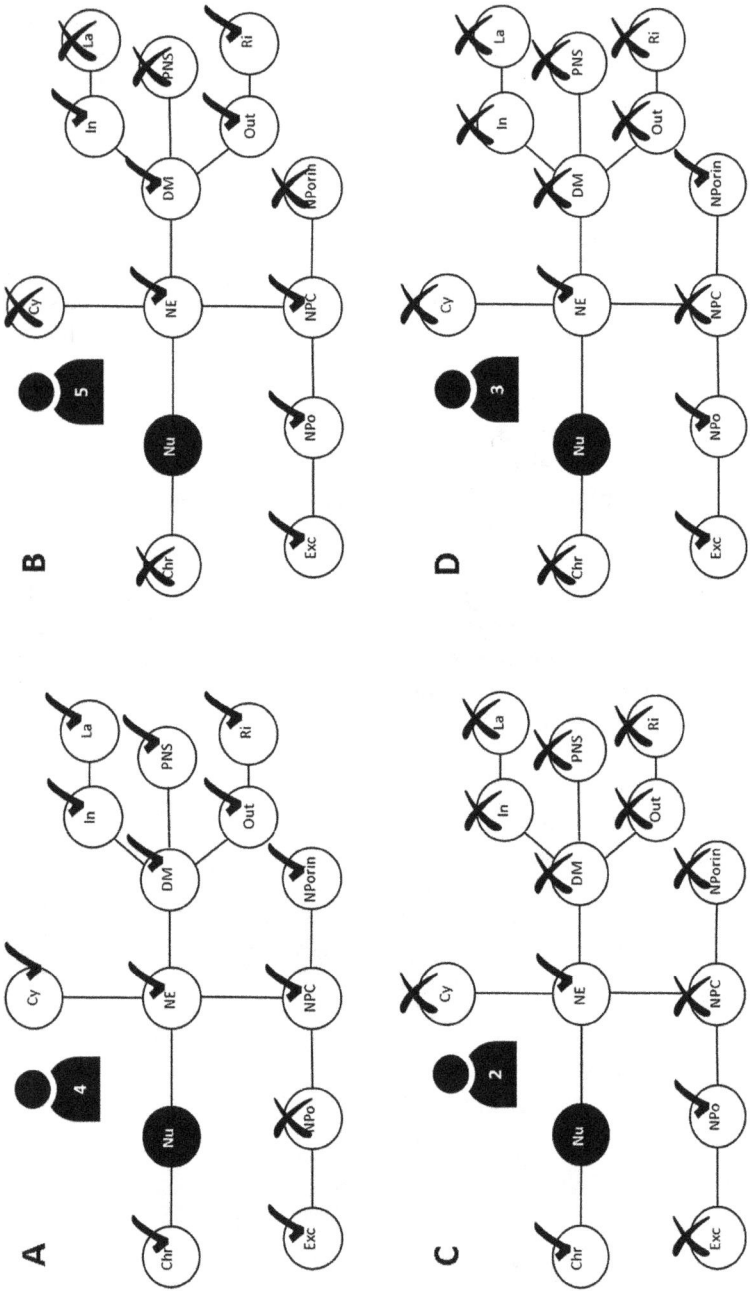

Figure 8.4 A–D Concept maps of students' biology vocabulary describing the nucleus of a eukaryotic cell.

mark represents a concept being used and an 'X' indicates that the student omitted the specific term. Nu = Nucleus; NE = Nuclear Envelope; Cy = Cytoplasm; DM = Double Membrane; In = Inner Membrane; Out = Outer Membrane; La = Lamin; Ri = Ribosome; PNS = Perinuclear Space; NPC = Nuclear Pore Complex; NPo = Nuclear Pore; NPorin = Nuclear Porins; Exc = Exchange; Chr = Chromatin/Chromosome). Student 5 applied fewer of the relevant terms to depict this cellular structure (Figure 8.4b), although his understanding and expression of the relations between the concepts was skilful and sound. In contrast, Students 1, 2 and 3 used a limited number of terms (some of which appeared in the question) and demonstrated significantly less comprehension of the concepts, despite the earlier developmental opportunity. Their understanding of how the concepts and components fit together and relate to each other, using power grammar, was also impeded (Figures 8.4c–d). These three students struggled to access the complex discipline-specific knowledge, or as Boughey (2002) argues, had problems with 'manipulating the forms of the additional language in a way that would allow them to receive and pass on the thoughts developed in the disciplines.' Valencia (2014) reasons that many students struggle to use their textbooks because 'the chapters are long and packed with specialized vocabulary; assumed background knowledge that students often don't have.' In contrast to Students 1 to 3, Student 6 was able to use many of the more complex terms (power words) as shown in Figure 8.5. Interestingly, this student is English-speaking (home language) but did not take Biology at school. So, although we witnessed her working hard to obtain the necessary power words (Biology vocabulary), her insight into how these concepts fit together

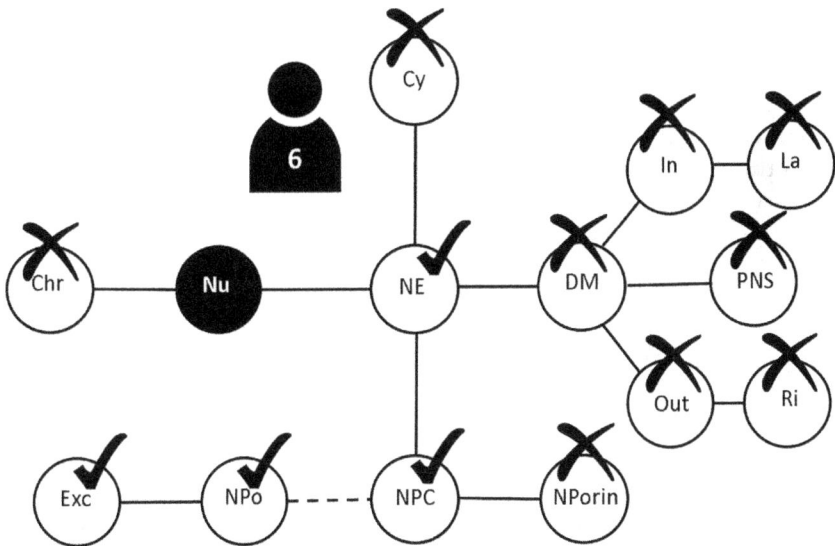

Figure 8.5 Concept map of Student 6's biology vocabulary describing the nucleus of a eukaryotic cell.

and relate to one another, needed more time for development (Figure 8.5), which restricted her ability to exhibit power composition. The aspect of discipline-specific scientific vocabulary, therefore, affects conceptual understanding, as well as the ability to communicate that understanding using proficient scientific discourse and power grammar. When students face science assignments or assessments, they need words, ranging in complexity, to firstly think about and process questions, ideas, possibilities and possible answers. Thereafter, they need a discipline-specific vocabulary (power words) to communicate their answers and ideas, either through verbal or written discourse. Dyasi (2006) therefore contends that 'words represent intelligence; acquiring the precise vocabulary and the associated meanings are key to successful scientific thinking and communication.' Our observations and analyses corroborate these arguments, which seems to be particularly true for students who were not instructed in English at school. It, therefore, appears that students who were instructed in English at school, despite having a different home language, have an advantage when it comes to navigating the gap between the school and first-year curricula and textbooks. These students have developed more skills to manipulate the forms of their additional language to express themselves and share their knowledge.

'Unpacking' of complex concepts

Another interesting finding was that the discourse of the students varied considerably with regards to the degree of explaining or 'unpacking' of complex concepts. In Figures 8.2 and 8.3, the complexity of meaning is represented by the respective segments of the stacked bars with the more complex meaning towards the top of the bars and simpler meaning towards the bottom of the bars. When one considers sections 6 and 7 on these stacked bars, these segments represent terms with relatively stronger complexity of meaning, which should be mastered by the students in this curriculum. Our results showed that the students were all able to use terms from both these desired categories. However, some students (Students 4, 5 and 6) excelled in using these terms with greater complexity appropriately. Students 4 and 5 further demonstrated a deep understanding of how these concepts relate to one another. An example from Student 4's discourse demonstrates that this student was able to use her Biology vocabulary (power words and grammar) to write a detailed description of the nuclear envelope of a eukaryotic cell:

> The membrane surrounding the nucleus is called the nuclear envelope. The nuclear envelope has two membranes. It has an inner membrane and an outer membrane and it contains an inner membrane space between the two membranes. The nuclear envelope also separates the nucleus from the cytoplasm and serves as a type of protection as it contains the delicate genetic information.
>
> (Student 4)

This student displayed excellent understanding and mastery of both the knowledge and scientific vocabulary using power words and grammar to describe this cell structure from various perspectives, and we regard this as an example of power composition. Even though she repeats herself to some extent, her detailed description would allow even a novice in the field to form a mental image of this structure. In contrast, despite Students 1, 2 and 3's discourse including some terms from these two high-end categories, they displayed significantly less understanding in their descriptions. For example:

...that will pass through the nuclear envelope (a membrane that covers the nucleus and allows for specific substances passage...).

(Student 1)

It [the nucleus] has a membrane called the nuclear envelope that encloses the nucleus' substances and structures as well as allows substances to pass through.

(Student 3)

It was also thought-provoking to notice how the students' descriptions of certain structures differed from the same descriptions in the school and first-year Biology textbooks. Even the less proficient students showed some development in their knowledge when compared to the school textbook, which described the nuclear envelope by saying:

The nucleus is surrounded by a double nuclear membrane with pores. The pores form the passage between the nucleus and cytoplasm of the cell.

(School Textbook)

Despite the school textbook being moderately dense in meaning, it is relatively low in volume and not very complex. The section we analyzed did not contain any words from the stronger categories, 6 and 7 (Technical; Compact; Properties and Technical; Conglomerate; Elements). We calculated that the text comprised approximately the same number of words from category 5 (Technical; Compact; Elements) as the first-year textbook, however, the text never reached the strongest levels of complexity in meaning in its descriptions of these concepts.

In contrast, the first-year textbook presented both complex and very dense meaning (ESD+). Moreover, there is significant epistemological condensation from the school to the first-year textbook. Condensation refers to the process of adding meaning, in this case to biological terms. Moreover, the descriptions of these terms in the first-year textbook were very concise, compared to the discourse of the proficient students. The students used complex words from both these desired categories (6 and 7), and some students (Students 4, 5 and 6) excelled in using words with strong complexity. The difference between the discourse of these three students and the first-year textbook can be ascribed to the extent of explaining and 'unpacking' done by them when describing the relevant structures. Student 4 used a similar number of complex words as the first-year textbook. However, she repeatedly used specific terms (e.g. 'nuclear envelope'), each time linking the term to a different aspect, information or point of view, revealing different perspectives. In contrast, the first-year Biology textbook uses such a term only once in its description of the same concept. Thus, the way Student 4 elaborated on and constructed the text made her discourse less dense and compact, and therefore more accessible, especially to novices, as she was 'unpacking' the condensed meaning gradually and systematically, shifting regularly between complex and simpler meaning. Referring back to the wording tool and our fourth category (Table 8.2), it is evident that this student often used phrases such as 'called the' and 'just like the,' which makes her written descriptions less compact, and in our opinion more accessible to novice readers. We, therefore, argue that Student 4 models the role of the lecturer in the teaching and learning process by 'unpacking' and 'repacking' the concepts in much detail, using a fair amount of repetition and pointing out various perspectives of the same concept (Figure 8.6; ★ = 'nuclear envelope'). This discourse is a perfect example of Maton's (2013) *semantic waves* where regular shifts can be seen between more complex and simpler meanings. Figure 8.6 shows how the discourse in the first-year textbook compares to that of Student 4 for the description of a specific cellular structure. Despite the textbook description displaying semantic waves, it is much more dense and compact. We, therefore, argue that the first-year textbook is semantically very dense, thereby failing to facilitate epistemological access for many students to this powerful knowledge. Lecturers, therefore, need to make an explicit effort to make this written discourse more accessible to students by 'unpacking' the complex meaning of concepts gradually and systematically (Mouton and Archer 2018) in the way Student 4 modelled.

Throughout our analysis, using the word-grouping tool, we found that proficient students often repeated themselves and frequently used phrases

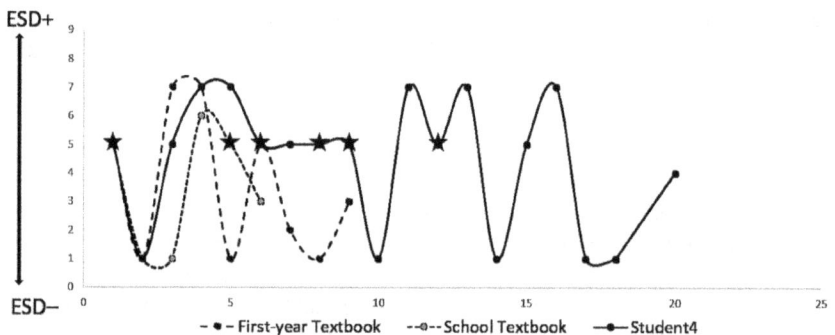

Figure 8.6 Semantic density profile of concept 1 component 2 from first-year and school textbooks, and from final summative assessment of student 4.

such as 'called the' and 'just like the,' that helps to 'unpack' meaning but also enacts accessibility to novice readers. For this study, we, therefore, used an adjusted version of the word-grouping tool with a fourth modification for our analysis that we named 'Defining/Relating' (Table 8.2). This is not the type of phrasing that is typically found in more formal texts such as textbooks. However, in the students' written text, these modifications tended to make the text less dense by being more specific (defining), or by relating preceding words to other concepts. Examples of such modifications would be 'membrane <u>called the</u> nuclear envelope,' or '<u>just like the</u> plasma membrane, the nuclear envelope prevents…'

Going forward

After reflecting on the results of this study, as well as studying literature showing that 'science talk' is cognitively less demanding than 'science writing' (Institute for Inquiry 2015), the project part of this study was amended for future cycles in three ways: Firstly, within their groups, the students will be given time to work through a given portion of the first-year textbook, discuss the content and make a list of the Biology vocabulary (power words) found in the text. Secondly, they will have to construct a concept map of the given cellular structure to include the Biology vocabulary. The concept maps are meant to promote the processing and synthesis of concept knowledge and to reduce the cognitive load, but also help students discover how each term relates to others in the bigger constellation of meaning. The concept maps will then be used as a basis for the structuring of the written discourse for the project. Thus, students will have time and opportunity to first 'unpack' the complex, dense meaning verbally and collaboratively, and thereafter in a visual concrete manner by constructing a concept map, before having to formulate 'high-stakes' scientific writing. Lastly, an online Biology dictionary has been compiled to help students understand and 'unpack' complex and compact terms by being able to quickly check the meaning of a term before using it in their discourse. In future studies, we plan to investigate the impact of these interventions.

Conclusion

This collaborative project was intended to make the literacy practices and 'genre' of Biology explicit to the first-year students through collaborative reading and writing activities (Jacobs 2007; Kirk 2019). Another objective was to provide students with an opportunity to engage in 'high-stakes reading' by using their Biology textbook and other literature, develop their Biology vocabulary and knowledge, followed by engaging in scientific writing by compiling a report, thus producing 'high-stakes writing.' The project was followed by a summative assessment, which presented another opportunity to showcase their mastery of the content, but also provided material for analyses of their writing skills. We believe that these activities brought some aspects to light: Firstly, the startling and underestimated difference between the school and first-year Biology textbooks in terms of volume, complexity and condensation of meaning, which would explain why so many students struggle to use the first-year textbook effectively. The gap between these two resources is substantial, and lecturers need to be made aware of this to assist students with this transition. Secondly, the variation in the proficiency and command of the students in terms of scientific vocabulary is noteworthy. Some students manage to gain and use a substantial volume of new scientific vocabulary (power words and grammar), while many others struggle to master the much higher volumes and accompanying complex meaning. These students need more time and opportunities to engage in using scientific discourse. Lecturers need to be aware of the wide variation in skills between students and attempt to support them by developing and sharpening these skills. Finally, the variation in skills among the students to manipulate the forms of their additional language in a way that would allow them to receive and pass on the knowledge they have developed in Biology, needs to be acknowledged. This study revealed how proficient students skilfully elucidate complex meaning by gradually 'unpacking and repacking' compact meaning from the textbook. We believe that this is also the role of the lecturer, to unpack and repack the complex, compact meaning for all students.

We believe that learning activities such as the one featured in this study, as well as the ones that will be included in future cycles as a result of these findings (e.g. the construction of concept maps), implicitly 'include' students to successfully engage with the discipline of Biology, and its values and ways of thinking. It contributes to the development of vital skills such as students' scientific discourse but also their identities as future scientists.

Acknowledgements

A sincere thank you to our academic literacies colleague's Mrs Y. Coetsee, Dr A. Binneman, Mrs E. George, Mrs E. Basson, Dr C. Fourie and Mrs F. Haroun for their valuable contributions to this project.

References

Ambitious Science Teaching. (2015) 'A discourse primer for science teachers.' http://ambitiousscienceteaching.org/wp-content/uploads/2014/09/Discourse-Primer.pdf

Boughey, C. (2002) "Naming' students' problems: An analysis of language-related discourses at a South African university,' *Teaching in Higher Education*, 7(3): 295–307.

Carrol, S., & Goodstein, D. (2009) 'Defining the scientific method,' *Nature Methods*, 6(4): 237.

Case, J., Marshall, D. and Grayson, D. (2013) 'Mind the gap: Science and engineering education at the secondary-tertiary interface,' *South African Journal of Science*, 109 (7/8): 1–5.

Chimbganda, A.B. (2000) 'Communication strategies used in the writing of answers in biology by ESL first year science students of the University of Botswana,' *English for Specific Purposes*, 19(4): 305–329.

Clarke, J. (2015) 'The value of understanding students' prior writing experience in teaching undergraduate science writing,' *Critical Studies in Teaching and Learning*, 3(1): 21–43.

Dyasi, H. M. (2006) 'Visions of inquiry: Science,' in R. Douglas, K. Worth, and W. Binder (eds) *Linking Science and Literacy in the K–8 Classroom*, NSTA Press.

Engle, R.A. and Conant, F.R. (2002) 'Guiding principles for fostering productive disciplinary engagement: Explaining an emergent argument in a community of learners classroom,' *Cognition and Instruction*, 20(4): 399–483.

Gee, J.P. (1996) *Social linguistics and literacies: Ideology in discourses*, Taylor and Francis.

Gee, J.P. (2005) *An introduction to discourse analysis: Theory and method*, Routledge.

Hurst, E. (2010) 'Language in Engineering and the built environment: Examining students' problems and impacts,' *Language*, 1986.

Institute for Inquiry. Exploratorium. (2015) 'Science writing: A tool for learning science and developing language,' www.exploratorium.edu.

Jacobs, C. (2007) 'Towards a critical understanding of the teaching of discipline-specific academic literacies: Making the tacit explicit,' *Journal of Education*, 41: 59–81.

Kelly-Laubscher, R.F. and Luckett, K. (2016) 'Differences in curriculum structure between high school and university biology: The implications for epistemological access,' *Journal of Biological Education*, 50(4): 425–441.

Kelly-Laubscher, R.F., Muna N. and van der Merwe, M. (2017) 'Using the research article as a model for teaching laboratory report writing provides opportunities for development of genre awareness and adoption of new literacy practices,' *English for Specific Purposes*, 48: 1–16.

Kelly-Laubscher, R.F. and van der Merwe, M. (2014) 'An intervention to improve academic literacies in a first year university biology course,' *Critical Studies in Teaching and Learning*, 2(2): 1–23.

Kirby, N.F. (2010) 'Foundation science student performance explained. Insights gained at UKZN,' https://www.assaf.org.za/files/2010/10/Kirby-Performance-explained.pdf (Accessed 28 November 2019).

Kirk, S. (2019) 'Knowledge-orientated perspectives on EAP: Some tools for thinking and practice,' paper presented at the Knowledge in EAP BALEAP Professional Issues Meeting (PIM) University of Northampton, 22 June 2019.

Lee, O. and Fradd, S.H. (2002) 'Instructional congruence to promote science learning and literacy development for linguistic diverse students,' in D.R. Lavoie and W.M. Roth (eds) *Models of Science Preparation. Science & Technology Education Library* (Vol. 13), Springer.

Marshall, D. and Case J.M. (2010) 'd/Discourse in the learning of physics: The design of an introductory physics curriculum,' *African Journal of Research in MST Education*, 14(2): 15–27.

Martin, J.R. (2013) 'Embedded literacy: Knowledge as meaning,' *Linguistics and Education*, 24: 23–37.

Maton, K. (2013) 'Making semantic waves: A key to cumulative knowledge building,' *Linguistics and Education*, 24: 8–22.

Maton, K. (2014a) *Knowledge and knowers: Towards a realist sociology of education*. Routledge.

Maton, K. (2014b) 'A TALL order? Legitimation Code Theory for academic language and learning,' *Journal of Academic Language and Learning*, 8 (3): A34–A48.

Maton, K. and Doran, Y.J. (2017) 'Semantic density: A translation device for revealing complexity of knowledge practices in discourse, part1 – wording,' *Onomázein. Revista semestral de lingüística, filogía y traducción, número especial SFL*, 2017: 46–76.

McNeill, K.L., Lizotte D.J. and Krajcik, J. (2005) 'Identifying teacher practices that support students' explanation in science,' paper presented at the *annual meeting of the American Educational Research Association*, April 2005, Montréal, Canada.

Mouton, M. (2019) 'A case for project based learning to enact semantic waves: Towards cumulative knowledge building,' *Journal of Biological Education*, 54(4): 363–380.

Mouton, M. and Archer, E. (2018) 'Legitimation Code Theory to facilitate the transition from high school to first-year biology,' *Journal of Biological Education*, 53(1): 2–20.

National Research Council. (2007) 'Teaching science as practice,' in R.A. Duschl, H.A. Schweingruber and A.W. Shouse (eds) *Taking science to school: Learning and teaching in grades K–8. Committee of Science Learning, Kindergarten through eighth grade*. The National Academies Press.

Skovholt, K. (2016) 'Establishing scientific discourse in classroom interaction teacher students' orientation to mundane versus technical talk in the school subject Norwegian,' *Scandinavian Journal of Educational Research*, 62(2): 229–244.

Valencia, S. (2014) 'When high school students struggle with textbook reading,' https://www.edutopia.org/blog/students-struggle-with-textbook-reading-sheila-valencia (Accessed 19 July 2018).

9 Using Autonomy to understand active teaching methods in undergraduate science classes

M. Faadiel Essop and Hanelie Adendorff

Introduction

Despite humankind entering the Fourth Industrial Revolution, higher educational institutions still largely rely on passive teaching and learning practices adopted more than a century ago. Here the lecturer becomes the 'sage on the stage' with students often required to passively acquire subject content and sometimes reduced to 'spectators' in terms of the learning process. However, the sole reliance on such teaching and learning practices is at odds with personality traits that characterize the current university student cohort. The Generation Z student (born: 1995–1999) typically exhibits a relatively limited attention span, a strong need for social interactions, a preference for critical thinking and problem-solving tasks, and a desire to express their opinions (Rothman 2014). The pervasive passive teaching and learning approach often also contradicts graduate attributes stipulated by many tertiary institutions. For example, at Stellenbosch University (Stellenbosch, South Africa) the intent is to produce graduates who display attributes such as an enquiring mind, critical and creative thinking, leadership and collaboration, problem-solving and an innovative outlook. Thus, it is clear that there currently exists an ideological and implementation gap between *actual* teaching and learning practices versus expectations and aspirations of both the university management/leadership and the wider student body. In order to begin to address this chasm, there is an increasing move towards the introduction of active teaching and learning practices into university lecture halls as this requires students to more deeply engage with subject content within the classroom setting (Goodman *et al.* 2018). This is in firm agreement with the 'pedagogy of engagement' (Smith *et al.* 2005) and problem-based learning as it is student-entered and allows for the successful interrogation and reflection of subject content by the students whilst at the same time applying critical thinking and problem-solving skills (Eberlein *et al.* 2008). Thus, it is a reasonable argument that science teachers should spend more time on how scientists *do* science (the process) and less on subject content to empower students to attain some of the skills highlighted earlier.

Focusing on the discipline of Physiology, a recent publication attempted to better define the meaning of the term 'Physiology,' both in terms of its

DOI: 10.4324/9781003055549-12

method and its specific object. Here, Lemoine and Pradeau (2018: 243) state that in terms of its method, Physiology is 'a quest to identify biological functions in organisms' and also 'a search for explanations based on biological functions.' They then continue to define it in terms of its specific object by stating that Physiology focuses on 'physiological and pathological phenomena, no longer as an explanation for clinical phenomena but described so that they can be explained (generally at a molecular level)' (*ibid.*: 243). In addition, Physiology focuses on '[t]he integrity of the organism, which may now also be accounted for by different, non-functional disciplines' (*ibid.*: 243); and 'all-encompassing phenomena, such as homeostasis, which, although not able to account for all biological phenomena, nevertheless provide useful and fruitful models for discovering new processes and understanding them' (*ibid.*: 243). Thus, Physiology is often regarded as a very basic discipline that deals with the functioning of the entire organism but that is integrated as a 'whole' by cross-talk that ranges from the molecular to the cellular, to the organ and the organismal level. This is a key factor that makes the subject particularly complex as students are required to understand both organ-related workings and then each one fits into the collective in order to ensure the overall health and well-being of the entire organism. For example, if an athlete exercises quite vigorously for an hour, then the impact of this external stressor should first be considered in terms of the different organ systems, for example, the brain, lungs, heart and muscles. Such information should thereafter be integrated to grasp the coordination and flow of information between different organ systems to ensure the athlete can indeed sustain the exercise regimen. This would equate to describing a physiological response to a vigorous exercise regimen. However, if the athlete is unable to cope with the external stressor, then a pathophysiological condition arises due to various problems both within such organ systems and in terms of their cross-talk and signalling. Students would also then be expected to understand, at a deep mechanistic level, how such pathophysiological situations may arise.

Based on personal experiences and conversations with colleagues in South Africa and abroad, Physiology students often grapple with large amounts of subject content to be covered in a module or course and therefore spend a significant degree of their time memorizing facts in order to pass tests and examinations. Although such memorization is useful, it also means that students often struggle to understand how different organs cross-talk to ensure an integrated and coordinated response(s) with the aim to re-establish homeostasis. Such 'surface' learning means that some students can display relatively limited problem-solving skills and lack deeper engagement with the subject content. There are numerous reasons that may help explain such behaviour patterns and learning approaches. For example, reporting on a United States faculty survey about possible sources of students' difficulty of learning Physiology, Michael (2007) highlights three factors that likely play a role in this case: the nature of the discipline (e.g. causal reasoning, integration), the way it is taught and what students bring to the task of learning

Physiology (e.g. learning equals memorization). These findings revealed that Physiology lecturers believed that it is the nature of the discipline and what the students bring to the learning of Physiology that renders it 'hard.' It is noteworthy that the *way* the subject is taught was viewed as less important when compared to the other two factors previously listed (Michael 2007).

The study also found that faculty's expectations were often too high, for example, faculty promoted information-overload which contributed to less desirable student approaches to learning, such as the focus on memorization. Here the author states that

> if physiology is hard because it requires the application of causal reasoning, and if students find it hard to apply causal reasoning to a physiological phenomenon, then we must first model this kind of reasoning for our students and give them ample opportunities to practice it while they receive appropriate feedback.
>
> (Michael 2007: 39)

It is therefore clear that such problems generally originate as a result of using only the passive mode of information transfer (from lecturer to student) that still predominates in most tertiary institutions (Wingfield and Black 2005). Whilst the traditional lecture still offers value, it is now well-established in the literature that active teaching approaches can ensure a greater engagement with subject content and also lead to improved student performances (Bonwell and Eison 1991; DiCarlo 2009; Michel *et al.* 2009; Goodman *et al.* 2018). Active teaching includes numerous activities for example quizzes, pair-share exercises and cooperative learning and is increasingly encouraged as a meaningful way to improve student performances and to also promote a more inclusive classroom environment (Michel *et al.* 2009; Goodman *et al.* 2018; Essop and Beselaar 2020; Essop 2020). Goodman (2018) and colleagues argue that 'meaningful learning has occurred when the students can solve appropriate problems with the mental models that they have built' (Goodman *et al.* 2018: 417). A comprehensive understanding of Physiology requires the ability to transfer knowledge gained in other disciplines, such as Chemistry and Physics, to the Physiology context and implies the ability 'to use disciplinary core principles' (Michael and McFarland 2011: 336). Thus, a greater ability to integrate or 'link what they had learned in other disciplines' would greatly enhance students' ability to learn 'some of the key concepts in physiology' (Goodman *et al.* 2018: 419). The integration of different aspects of the organism's Physiology (including concepts from other disciplines) into a 'whole' together with an increased understanding of cross-talk is particularly crucial when faced with problems that transcend specific topics covered in the curriculum, as is the case with most real-world problems. The aim of teaching students how to solve real-world problems (Klegeris and Hurren 2011) is thus often linked to a call for interdisciplinary collaboration aimed at integrating knowledges from various 'disciplines, rules, concepts, strategies, and skills' (Foshay and Kirkley 1998: 1). The assumption is also made that acquisition of the requisite knowledge means that students will be able to apply it in real-world problem situations (Bransford *et al.* 1986). However, such integration is unfortunately often hampered by knowledge blindness that can obscure how the different knowledge practices interact (Maton and Howard 2018). Furthermore, problem-solving is also seldom taught directly (Snyder and Snyder 2008).

Not surprisingly, the use of only traditional teaching methods is associated with poorer critical reasoning and problem-solving outcomes (Amin *et al.* 2017: 181). Priemer *et al.* (2020) point to a 'practice turn' in science education, marked by a greater focus on problem-solving, especially in domains that integrate 'content, procedural, and epistemic knowledge' (*ibid.*: 106). In a paper examining cumulative learning in Chemistry, Rootman-le Grange and Blackie (2018) argue that Chemistry education seems to focus on decontextualized conceptual learning or learning 'within a particular learning context (for example, practicals or organic chemistry)' (*ibid.*: 485). They postulate that trying to 'link conceptual development in a single context with the construction of knowledge' requires a 'leap from the ultra-specific grasp of this one concept to how this item is stored in long-term memory' (*ibid.*: 485). Building on their argument, we can conclude that the decontextualized siloed learning, often typical in science, would also not be conducive to the knowledge integration required for interdisciplinary fields such as Physiology where integration would similarly require students to 'move from the minutiae of the particular concept to the joining of this concept to the bigger picture' (Rootman-le Grange and Blackie 2018: 485).

By contrast, active and collaborative methods such as peer instruction and case-based learning (CBL), that reflect real-life situations where the lecturer acts as a facilitator and 'learning assistant' to the students are emerging as more effective alternatives (Kamran *et al.* 2011: 103; Rehan *et al.* 2016). Here, the role of the instructor is to help create an environment that can 'stimulate students' thinking and provide a situation for the students to analyze an authentic problem through the application of concepts and facts' (Amin *et al.* 2017: 181).

Problem-based learning (PBL) is another teaching approach put forward to help improve knowledge integration and problem-solving. In both PBL and CBL students 'critically analy[ze] contextualized (authentic) problems posed to them in a collaborative (group) setting' (Klegeris and Hurren 2011: 408). Facilitators exert slightly more control with CBL (compared to PBL) in that they typically employ guiding questions to ensure that learners do not stray too far from the original objective. In both approaches, the question is presented as a learning stimulus at the start of a course, section, topic or class (Essop 2020). Our 'CSI-type' activity is an example of a limited PBL approach where it functioned as 'a supplement to standard didactic lectures' (Klegeris and Hurren 2011: 408). This exercise is based on the well-known CSI television series where forensic investigators employ their scientific knowledge, laboratory tests, analytical and problem-solving skills to assess clues in order to help solve crimes in the United States. However, the implementation of such methods can vary from courses that are built around it to cases that include only some elements of it.

Despite broad agreement on the value and increased uptake of active teaching and learning methodologies, comparing the efficacy of the various forms of this approach in order to identify those most suited for teaching problem-solving in science courses such as Physiology remains a challenge (Goodman *et al.* 2018). This study, therefore, examined examples of in-house developed undergraduate Physiology problem-solving activities employed within the Department of Physiological Sciences at Stellenbosch University that aimed to foster critical reasoning and problem-solving skills.

Maton (2014) argues that most approaches to education suffer from 'knowledge blindness' that obscures an understanding of what happens in knowledge practices. A key example of this is found in interdisciplinary fields where it can obscure the role different fields play in knowledge-building. Physiology, drawing on various other scientific disciplines would be one such example. Problem-solving, especially as related to real-life cases, in Physiology would draw on even more diverse fields, including communication, problem-solving approaches and information literacy. Thus, bringing the knowledge practice in this case into focus would require being able to assess how these different elements relate in practice. The Autonomy dimension of Legitimation Code Theory (LCT) affords us a way of doing this. Through the use of *autonomy codes*, we can investigate how the different knowledge areas interact, highlighting the knowledge practice underpinning problem-solving in the context of Physiology. Gaining this understanding can help lecturers choose the most appropriate active teaching methods for integrating different knowledges required for real-world problem-solving.

Goodman *et al.* (2018) argue that 'meaningful learning [in Physiology] has occurred when the students can solve appropriate problems.' In this chapter, we are interested in comparing the value of different active teaching methods that can lead to such 'meaningful learning' manifested in the ability to solve real-life problems. Our focus will be on examples of active teaching methods used in Physiology.

Teaching methods

Our interest in this chapter is in the value of active teaching methods, such as PBL, for helping students apply physiological content to real-life or making sense of life-like cases through physiological concepts. Our focus will thus not be on the teaching of Physiology content but on how the real-life examples allow for the kind of knowledge integration required for problem-solving in Physiology.

Active teaching and learning methods

For the purpose of this chapter, we considered three active teaching methods, one of which was taken from an online repository at the National Center for Case Study Teaching in Science (NCCTS) at the University of Buffalo for use across disciplines and year levels. This PBL-type example was selected based on the NCCTS resource being recommended for use in teaching students problem-solving in Physiology specifically (Goodman *et al.* 2018).

The other two were developed in-house and used during the cardiovascular Physiology component of a final-year semester module for the Human Life Sciences stream of the BSc degree offered by the Faculty of Science at Stellenbosch University attended by approximately 200 students.

NCCTS example: A Can of Bull

The active teaching methods suggested by Goodman *et al.* (2018) include a reference to the repository at the NCCSTS (n.d.) at the University of Buffalo. The website suggests that such case studies aimed at making science relevant can be used in a variety of ways and contexts. We selected one of their Physiology examples, a case study focusing on energy drinks, called 'A Can of Bull' (Science Cases n.d.). This case requires students to analyze popular energy drinks in order to judge validity of the related marketing claims pertaining to their nutritional value. This problem case involves using knowledge from fields such as Biochemistry and writing an analysis for a popular magazine.

In-house method 1: Burning Questions

For this activity, a case study is displayed on screen during lecture time with the lecturer then recruiting volunteers (two to three students) to tackle it. The case study is designed to ensure that it will test the student's problem-solving abilities, enhance critical thinking and promote integration of various concepts. The student volunteers usually have around two days to research the question and are required to present their findings to the class (using three to four slides) during the next lecture. The students are encouraged to make contact with the lecturer to help clarify any difficulties they may experience, to provide them with relevant literature and to also help gain a more integrated understanding of the various physiological concepts. After their classroom presentation, the lecturer will provide constructive feedback regarding the strengths and weaknesses of their response. Such feedback is shared with the entire class. In parallel, the rest of the class is also encouraged to tackle the Burning Question(s) and to submit a short, written response for lecturer feedback at the start of the next class. Their responses should be submitted before the short presentation done by the student volunteers. It is important to note that the entire exercise is done on a voluntary basis and that no marks are on offer.

In-house method 2: Running Questions

For the Running Question, students are initially faced with a crime scene together with related evidence. This Running Question continues over several lectures, and additional evidence emerges at regular intervals. During this process, students in class are regularly engaged for their inputs and insights whereafter summaries of what is actually happening are provided by the lecturer.

Analytical framework: The Autonomy dimension of LCT

LCT is a toolkit for making explicit the organizing principles underlying knowledge practices and their contexts (Maton 2014). This framework allows for research to get beneath the surface features of empirical situations and to explore their organizing principles or 'codes.' LCT currently comprises four 'dimensions' or sets of concepts that reveal different kinds of organizing principles: Autonomy, Semantics, Specialization and Temporality. For this study, we employed the dimension of Autonomy, which explores knowledge practices in terms of *positional autonomy* and *relational autonomy* (see Maton and Howard 2018, 2021a, 2021b).

Positional autonomy (PA) is concerned with how insulated constituents or elements of one knowledge practice are from those in other knowledge practices. It is focused on where things come from, in other words, what is considered as part of or 'inside' a specific knowledge practice and what is not. Elements or constituents that are considered to be from 'inside' a practice are characterized by stronger positional autonomy (PA+) and those considered 'outside' are characterized by weaker positional autonomy (PA−).

Relational autonomy (RA) concerns the relations between principles from within a specific context and those from other contexts. In teaching and learning contexts it is often interpreted as the purpose of the learning task, thus distinguishing between purposes that are considered to be part of or 'inside' a specific practice and those that are not. Purposes that are considered to be 'inside' a practice are defined as having stronger relational autonomy (RA+) than those considered from 'outside' (RA−).

Autonomy in our example

One of the first steps in an autonomy analysis is to define the *target* of the study, in other words, the constituents (PA+) and purposes (RA+) that are viewed by agents as 'inside' or constitutive of the practice. Our interest in this chapter is in examining active learning methods, based on simulated real-life problem cases, in terms of their value for helping students apply physiological concepts in practical scenarios. By defining 'making sense of real-life physiological cases' with the use of 'relevant physiological concepts' as the *target* of our study, we can identify the physiological specifics of the case and the physiological concepts and methods pertaining to it as being of stronger positional autonomy (PA+) and elements outside of the problem such as the biochemical concepts or writing skills as being of a weaker positional autonomy (PA−). Similarly, we can define making sense of the case and explaining the case as having stronger relational autonomy (RA+). All other purposes, such as making sense of other scientific information pertaining to the case or communicating the results are defined as having weaker relational autonomy (RA–).

PA and RA can vary independently, tracing an infinite number of strengths. Their relative strengths can be represented on a Cartesian plane. Figure 9.1 shows the *autonomy plane*, on which there are four principal modalities:

Sovereign codes arise when both PA and RA are relatively strong (PA+, RA+). In our case, this translates to using physiological concepts (PA+) to explain the real-world problem information (RA+) thus learning to apply physiological concepts beyond the confines of the classroom (see Figure 9.2).

Exotic codes (PA–, RA–) result when information from elsewhere (not Physiology) is used for extrinsic purposes (not explaining the real-world problem case). An example of this would be using Chemistry concepts to Chemistry concepts to write an article on the usefulness of energy drinks.

In *projected codes* (PA+, RA–) materials and practices from inside (PA+) are used for extrinsic purposes (RA–), i.e. when physiological concepts are used to explain something other than the real-world situation – for example, using the reading on the blood sugar meter in the case to explain a related physiological concept.

In *introjected codes* (PA–, RA+) materials and practices from outside are used for intrinsic purposes; in other words, something other than Physiology (PA–) is used to explain the real-world problem case (RA+) – for example, drawing on Biology to judge energy references in the marketing claims. Figure 9.2 shows these codes unpacked in a simple form for this study.

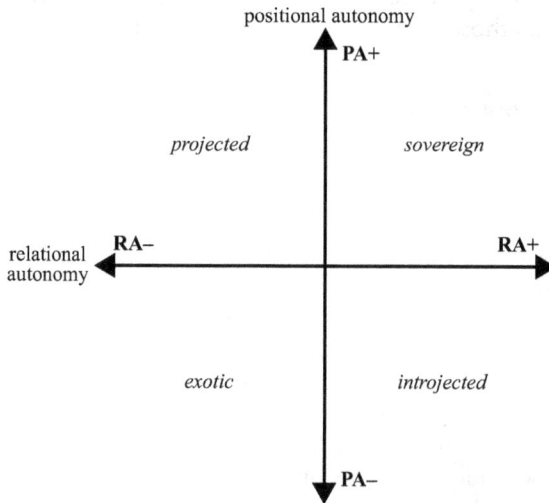

Figure 9.1 The autonomy plane (Maton and Howard 2018: 6).

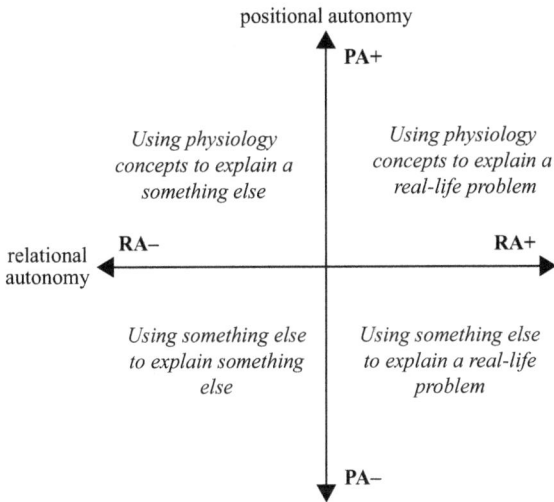

Figure 9.2 Autonomy plane for active teaching methods in physiology.

Autonomy pathways

The *autonomy plane* affords us with the opportunity to plot the manner in which the autonomy codes change over time in this context. Since both PA and RA can vary over a continuum of infinite strengths both can take an infinite number of positions in practice. When the aim of a task is to help students integrate knowledge from different sources or disciplines into a coherent whole and apply physiological concepts to explain real-world problems, we would expect to see the task reflecting this. Moreover, since applying physiological concepts in this way may require refining one's understanding of related scientific concepts, we may expect the task also to require the use of supplied information for a variety of purposes, including the learning of other scientific concepts. We can trace *autonomy pathways* (Maton and Howard 2018, 2021a, 2021b) or how activities move around different autonomy codes. There are an infinite number of possible pathways, of which Maton and Howard (2018) highlight four:

1. *Stays* when the activity stays within one code.
2. *One-way trips* where the activity starts in one code and ends in another.
3. *Return trips* (see Figure 9.3) which start in one code, then move to another before returning to the code which they started in. (This is the simplest form of an *autonomy tour*.)
4. *Autonomy tours* which start anywhere and can travel through one or more codes before returning to their code of origin.

Autonomy thus allows the knowledge practice of problem-solving in Physiology 'to be seen as an object of study in [its] own right' (Maton and Howard 2018: 8). Using the autonomy, plane it is possible to trace the steps

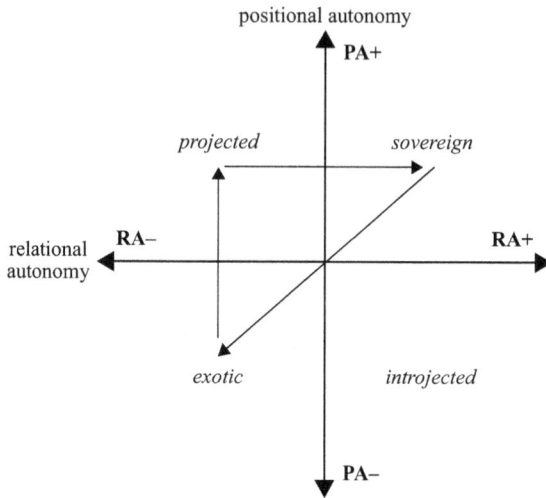

Figure 9.3 Example of an autonomy tour (Maton and Howard 2018: 9).

in the process or task as *pathways*. By plotting the codes related to each task step we can gain a better understanding of the underpinning rules of the game and how knowledge from different fields is integrated in this case. This also allows us on a more general level to observe how different kinds of information come together and are repurposed for a specific goal.

The second step in this kind of analysis involves the creation of what LCT terms a *translation device* or description of the ways in which the codes were enacted in analysis of the data. Translation devices are tools that allow 'translation between theory and practice' (Maton 2016: 243). In order to create this translation device, we followed a process similar to that described by Maton and Chen (2016). After spending some time in the literature familiarizing ourselves with the nature of the active teaching strategies we were to study, we started by defining the content and aims of the active teaching methods as our target. The strongest PA in these examples was thus defined as the physiological information contained in the real-world scenarios in the examples with theoretical physiological concepts and concepts from other sciences defining the next two levels, respectively, and content from contexts other than science representing the weakest PA. This was tested and refined multiple times by applying it to the practice, to finally yield the device in Table 9.1.

For each of the active teaching methods, the primary aim was to learn how to apply Physiology content to a real-life context. This aim would thus represent the strongest relational autonomy (RA++) with making sense of, or explaining, Physiology relevant to the case being of slightly weaker relational autonomy (RA+). Making sense of other scientific concepts required for understanding the relevant Physiology would be even weaker relational autonomy (RA–). Incorporating this into our data eventually resulted in the translation device in Table 9.2.

Table 9.1 Positional autonomy – a translation device

PA	Description	Examples from data
PA++	Physiological elements inherent in the real-life information supplied in the task or case	Energy drink marketing claims, i.e. 'contains special supplements to immediately enhance mental and physical efficiency'
PA+	Theoretical Physiology concepts relevant to the task or case	Physiological action of various components of energy drinks
PA–	Scientific information and concepts pertaining to the task or case	Chemical composition of energy drinks
PA– –	Other information pertaining to the task or case	Writing skills

Table 9.2 Relational autonomy – a translation device

RA	Learning to:	Examples from data
RA++	explain/offer a solution for **real-world physiological scenario/observation** in physiological terms	Explain when energy drink might be useful
RA+	make sense of or explain **theoretical physiological concepts** relevant to the case	Explain physiological action of chemical compounds in energy drinks
RA–	make sense of or explain general **science concepts** relevant to the case	Categorize chemical compounds in energy drinks
RA– –	**use other skills** that are relevant to the problem	Writing for a popular magazine

Examples of active teaching methods

Example 1: A Can of Bull

In this case study, which focuses on 'large biomolecules, nutrition, and product analysis,' students are asked to: (1) scientifically support or refute the nutritional marketing claims of these products, (2) determine the conditions under which such energy drinks might be useful and (3) translate this information into an article for a popular magazine. (Science Cases n.d.). In order to do this, the problem is broken down into five steps, starting with being told (1) that they will need to look at a chemical description and categorization of components of various popular energy drinks and (2) the physiological role of these components in the human body before judging the marketing claims, suggesting conditions for use and writing the article.

The case (Science Cases n.d.: 1) opens with a narrative about Rhonda 'who had landed the job of her dreams as a writer for Runners' World magazine. We are informed that she 'had excelled in cross country … had been a consistent runner, participating in local races and those assigned to her for her job.' Then '[a]s if reading her mind, her boss Charley walked in just then with a can of XS Citrus Blast® in one hand and a list of several other energy drinks in the other' (available at https://sciencecases.lib.buffalo.edu/files/energy_drinks.pdf).

In this introduction, aimed at piquing the interest of the students, we see the case opening as an *exotic code*. We are given information about Rhonda (not pertaining to the case or science related to the case, PA– –) for reasons other than explaining the case or making sense of the science related to it (RA– –).

The case description follows with an exchange in which Rhonda and Charley discuss what they know about energy drinks, with Rhonda saying that 'it seems they're primarily used by athletes to provide some "fuel" as they practice and compete. Other people use them more casually as a way to become "energized". That's about all I know' (Science Cases n.d.: 1). With this, the case moves into a *projected code* using case-specific information (energy drinks) and what they are used for (PA+) to draw students into the case (RA–).

Charley then tells Rhonda 'to find out what each of the ingredients in these drinks is and what it does for a runner or for a non-athlete.' She is cautioned to accurately 'determine what each component really does for the body, not what the marketers want you to believe it does' (Science Cases n.d.: 1). In order to help her with this, she is provided with the calorie value, marketing claims, ingredients, nutritional facts and biochemical information for each energy drink. She is given a list of questions, starting with the following:

Question 1: 'When we say that something gives us "energy," what does that mean? What is a biological definition of energy?' (*ibid.*: 4)

This question requires the use of scientific information (PA–) to make sense of the relevant science (RA–), thus starting the task as an *exotic code*.

Question 2: 'What is the nature (sugar, amino acid, vitamin, etc.) of each ingredient listed on the cans?' (*ibid.*: 4)

This question still demands the use of scientific information pertaining to the contents (PA–), but this time to help make sense of an aspect of the problem (determining the contents of the energy drinks, RA+), thus moving the task to an *introjected code*.

Question 3: 'What is the physiological role of each in the human body?' (*ibid.*: 4)

In this step, students use physiological information pertaining to the contents (PA+) to make sense of part of the problem (what this does to the body; RA+), thus transforming the task into a *sovereign code*.

Question 4: 'Which ingredients provide energy and how do they do that? Which ingredients contribute to body repair, i.e., which help build or rebuild muscle tissue?' (*ibid.*: 4)

Both of these questions, like question 1, require the use of product information (PA++) to answer a scientific question not directly related to the problem (RA−), thus moving back into a *projected code*.

- In the next four questions in the case (Science Cases n.d.), the task takes students through another similar cycle: *exotic code*, since they use science, the concept of a metabolic energy source (RA−), to explain a physiological concept, the perception of increased energy (PA−), when looking at how energy drinks that do 'not have a metabolic energy source provide the perception of increased energy' (*ibid.*: 10).
- *Sovereign code*, using theoretical Physiology concepts, the physiological action of different ingredients (PA+), to explain a part of the case, usefulness of the energy drinks (RA+), when explaining how 'the ingredients in these drinks [might be] helpful to someone expending a lot of energy' (*ibid.*: 10).
- *Sovereign code*, when using the acquired physiological knowledge (PA+) to substantiate the marketing claim that this is an 'energy drink' (RA++).
- *Projected code*, when using the information provided about the energy drinks, claims about what they can do (PA+), to explain physiological concepts, how the drinks will affect different people (RA−).

There are then three more questions, all projected codes using Physiology to explain something other than the case – e.g. the 'normal physiological response to increased intake of sugars,' 'sugar high' (Science Cases n.d.: 10) and the relationship between energy and sleep. After which students are asked to determine if the product claims are legitimate and to give reasons, thus locating this step of the task in a sovereign code since it requires them to use relevant physiological knowledge to explain the case.

Following the steps related to the set of questions, we see that this case study provides an autonomy pathway, travelling through various autonomy codes. This autonomy pathway starts in an exotic code and ends in a sovereign code (see Figure 9.4), with various stops along the way. What is important here is not the order of the stops (or task steps) in the pathway but the fact that students spent time in each code. This enhances the idea that problem-solving in real-life cases or explaining such scenarios draws on more than just Physiology concepts. The final output for this task, not coded above, asks students to communicate their findings by writing an article for a popular magazine taking them into an exotic code. This step is asking them to use something from outside the target (writing skills) to communicate their physiological knowledge. We can thus understand why they might struggle with this last element because these skills may not have been developed as part of their Physiology course, and the students may not see the value of it.

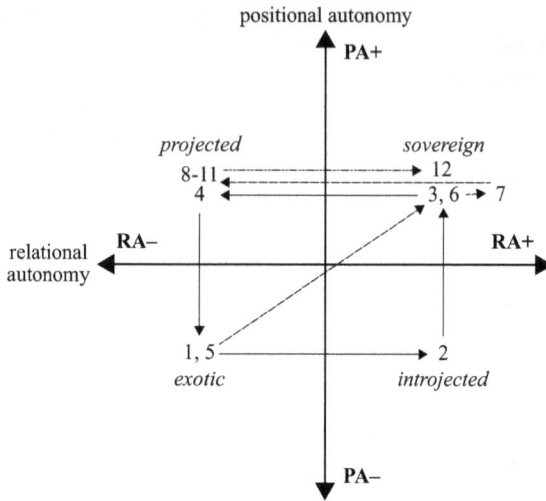

Figure 9.4 The autonomy tour traced by the first four task steps in a Can of Bull.

Example 2: A Burning Question

As mentioned earlier, this activity is usually presented as a real-life scenario, offered as an unexpected activity sometime during the lecture. The lecturer would read the slide (Figure 9.5) to the class and then ask for volunteers to prepare a response:

The invitation to participate is usually followed by a slide (see text below) with a reminder to a specific problem-solving approach, which they know as the 'HWW' principle:

> During this time, I inform the class of my principle which I call HWW: How does it happen, what, why? Why is it important? Explain it to lay persons … and I quote Albert Einstein to them: if you cannot explain it to a six-year-old then you don't understand it yourself. You cannot hide behind the jargon: I'm sorry this is complicated … no. You've got to really know it well to explain to bystanders. … Remember the theme is fun. So, you've got to bring the content, but in a way that is also fun.
>
> (Essop 2019: 16:25)

This activity starts as a projected code, offering physiological information pertaining to the case – age, medical information – in order to set the scene for students and draw them into the case (RA–). The first task (step 2 in Figure 9.6), which asks them to focus on what was learned in class to explain what is wrong with her, is a sovereign code: apply Physiology concepts from their course (PA+) to a real-life situation in order to explain what could be wrong (RA++).

The second task needs to be unpacked into two steps: (1) suggesting a course of action; and (2) providing scientific (here taken to mean

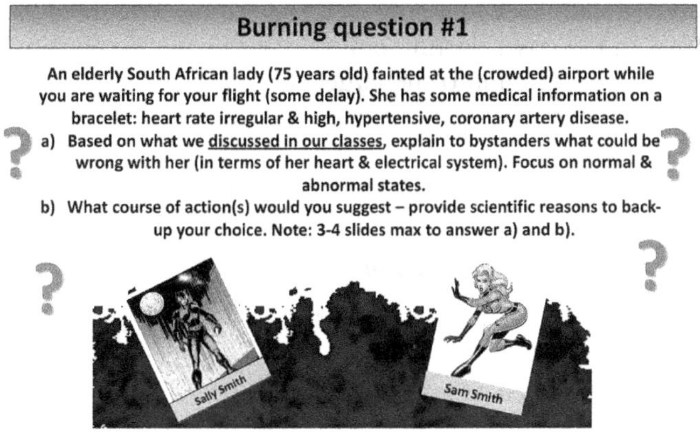

Figure 9.5 Example of a Burning Question.

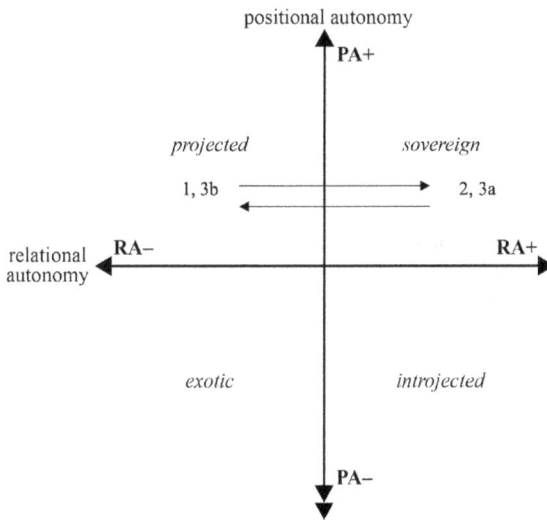

Figure 9.6 The return trip traced by the steps in the Burning Question.

physiological) reasons (3a and 3b in Figure 9.6). The first of these would keep them in a sovereign code, provided they base their advice (RA+) on physiological knowledge (PA+) and not on general science or any other knowledge. The second part proving scientific reasons requires the students to explain the Physiology (RA–) using physiological concepts (PA+), so this would take them back into a projected code, using Physiology concepts to explain their advice. It is important to recognize the distinction here between physiological concepts and the particular case. Here the focus is on the physiological concepts.

The steps in this task trace a *return trip*, from a projected code (using the case to excite them) through a sovereign code (applying physiological

M. Faadiel Essop and Hanelie Adendorff

concepts to the case) back to a projected code (using Physiology to explain their advice). In this case, starting outside the target sovereign code in the set-up, students are taken into the target sovereign code with the first step and then back out again when returning to more conceptual Physiology learning. This task, therefore, serves both to help them apply Physiology concepts and to strengthen their understanding of Physiology concepts.

Example 3: A Running Question

This type of activity is usually introduced at the start of a series of lectures, similar to the scenario where PBL is used as 'a supplement to standard didactic lectures' (Klegeris and Hurren 2011: 408). It has been suggested that students who generate responses to new problems before receiving teaching taught on the topics display deeper understanding than those who tried the problem after receiving instruction (Hmelo-Silver *et al.* 2018: 215). The problem-solving process can be left open-ended or students can be guided, by providing them with a conceptual model for approaching the problem (Jonassen 1997). Our example combines these two strategies by presenting the students with the problem prior to instruction and modelling the steps in the process of applying Physiology concepts to the case.

The activity starts with the lecturer introducing the slide text as follows:

> How are you guys feeling this morning? I just went for a lovely walk in the Jonkershoek mountains in Stellenbosch. Gorgeous day. Oh, we've got a problem there. So, we've found this Finnish girl has gone missing and we need Stellenbosch University students to assist us. And then, progressively I will give them clues.
>
> (Essop 2019: 27:30)

The next slide that is displayed then offers a number of clues (see Figure 9.7):

- An image of a banana peel and a half-eaten sandwich.
- [speech bubble from cartoon character] 'Maybe the girl had a little snack before disappearing?'
- An image of a meter with an 18.8 reading on it.
- [speech bubble from cartoon character] 'I know! My grandma uses it to check her blood sugar!'

At this point, the lecturer stops and asks the students to consider the information, before displaying the next slide (Figure 9.8), with one of the characters asking, 'What does a reading of 18.8 mean? This may help us.' The class is then led to consider whether the girl had something to eat and to consider the reading on the meter.

The introduction of the task starts as an exotic code since it sets up the scenario but without offering any physiological information (PA−) to pique

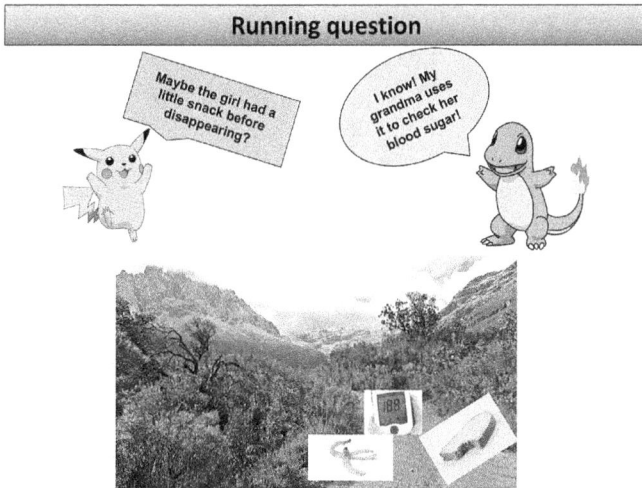

Figure 9.7 Running Question – first information slide.

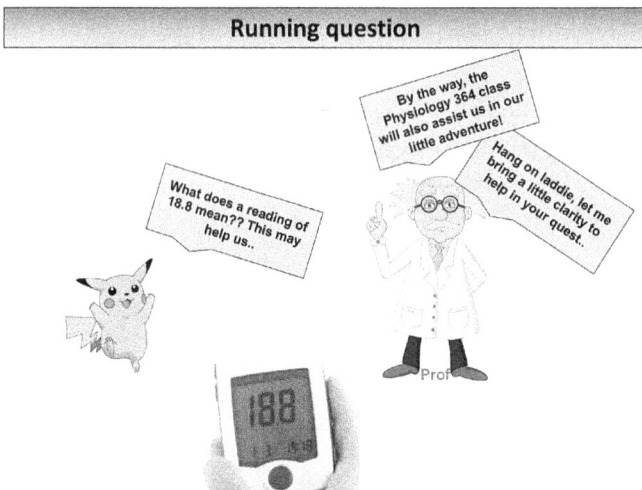

Figure 9.8 Running Question – second information slide.

student interest (RA–). The slide in Figure 9.7 displayed after the opening offers some physiological information/clues, thus strengthening the positional autonomy (PA+) and moving the task into a projected code. However, given that the aim still only is to get students engaged (RA–) they are not asked to make sense of the situation yet.

During the discussion, the class is directed to the question about the meaning of the information (Figure 9.8), which implies making sense of/ explaining a part of the scenario, i.e. the reading and the possibility that the girl had a banana before disappearing (RA+) using physiological concepts

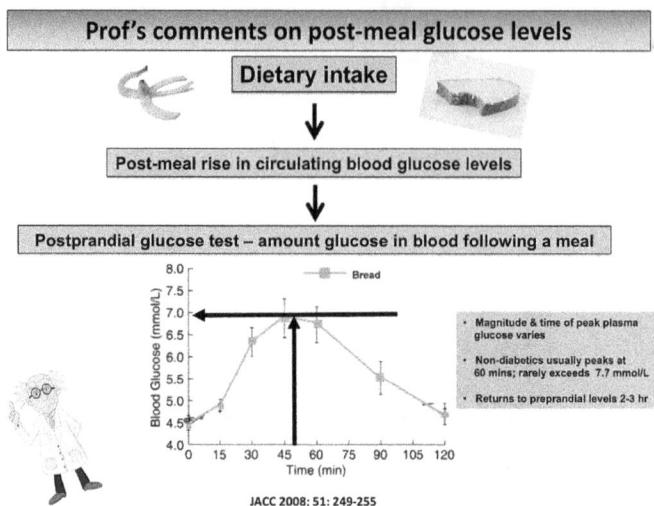

Figure 9.9 Running Question – theory slide.

(PA+), thus moving the task into a sovereign code. After the discussion, the lecturer brings in relevant physiological concepts (Figure 9.9) on post-meal glucose levels takes the discussion into a projected code by using Physiology information (PA+) to teach Physiology (RA−).

In the next step, the cartoon character appears again, asking 'So, the value of 18.8 mmol/L suggests quite a high value! Any idea what this would mean?' Students are also told that the girl ate 30–60 minutes before she disappeared.

In a new slide, students are given a new clue:

- [offered by lecturer] 'Just got a call from the lab – her insulin levels are very low, virtually not detectable.'
- [cartoon character responds] 'This seems like a good clue, how does insulin fit in though?'

This clue brings in new Physiology (PA+) information, which they can use to make sense of a part of the case, the relevance of the clues (RA+), moving the action back into a sovereign code. As before, students are asked to relate this information (the reading and laboratory report) to Physiology content (the role of insulin) thereby 'mapping the problem onto prior knowledge' (Jonassen, 1997: 70) (RA+). This is followed by the lecturer explaining more theoretical concepts, using Physiology content to teach Physiology, thereby moving the action to a projected code once more.

This return trip, embedded in the bigger tour, started with them making sense of the reading using Physiology (sovereign code) and then learning more Physiology (projected code), before making sense of the newly supplied information, using Physiology concepts (sovereign code) and dipping

Table 9.3 Running Question 'autonomy tour' with multiple return trips

	Activity	Code	Task
Problem statement	Problem description	exotic and/or projected	Supply authentic real-life information to pique interest
Repeated cycles of adding and relating problem information	Making sense of the problem situation with the use of Physiology	sovereign	Demonstrate/ask students to relate **problem information** to the Physiology concepts – what does this mean? (Figure 9.8)
	Making sense of the Physiology related to the problem and information	projected	Use Physiology concepts and problem information to deepen Physiology understanding – let's look at the Physiology (Figure 9.4)
Problem conclusion	Problem solution	sovereign	Using Physiology to explain the case

back into Physiology concepts to deepen their understanding of Physiology (projected code), is repeated again when histological slides are introduced as evidence. After three such cycles, the problem is solved (the case is explained and the girl is 'found') (RA+) using Physiology concepts (PA+) to make sense of the problem, ending in a sovereign code. This autonomy pathway is summarized in Table 9.3, highlighting the multiple return trips between the start and conclusion of the tour.

Discussion

This analysis highlights key steps in the process of helping students apply Physiology concepts to an authentic scenario. By repeating the same cycle multiple times, the lecturer is helping students see the relevance and application of the physiological concepts, whilst deepening their conceptual understanding with each move. In each case, the lecturer starts with making sense of the information provided, using Physiology concepts, followed by deepening Physiology understanding. This is repeated a number of times as new problem information comes to light before applying this Physiology understanding to finally solve the problem in the final step. We see in this an example of a very deliberate process of helping students interpret case information with the use of physiological concepts.

When addressing real-life problems in an interdisciplinary field such as Physiology we would expect to see some travel across the autonomy plane.

Indeed, the three methods we looked at traced different autonomy path-ways, with varying amounts of travel across the plane. Using autonomy anal-ysis in this way can help us reveal which teaching methods might be better at facilitating the teaching of problem-solving in Physiology contexts.

It is important to note that with all these examples we considered the design of the active teaching method only. We did not interrogate the actual student experience or interpretation of experience. In the final example, we can trace anticipated student involvement through what the lecturer is doing, but we were still looking at the design only. Where students do the problem-solving themselves, their approaches might well trace different and poten-tially more complex autonomy tours. This would also be an interesting topic for a future study on the efficacy of different active teaching methods in Physiology.

Conclusions

In this chapter, we set out to assess the ability of Autonomy from LCT to compare different active teaching methods. The three examples provide us with insight into the logic underpinning the process of learning Physiology through authentic or real-life problem cases by highlighting how real-life information and Physiology concepts interact during the use of these active teaching methods. It is clear that real-life problems offer an interesting way to help students integrate and contextualize Physiology knowledge but that course designers need to be mindful of the underlying aim of the project. Using autonomy analysis in this study suggests a useful and powerful new means of designing and testing active teaching methods aimed at integrating knowledge from different sources in interdisciplinary courses.

References

Amin, A.M., Corebima, A.D., Zubaidah, S. and Mahanal, S. (2017) 'The critical thinking skills profile of preservice biology teachers in animal physiology,' *3rd International Conference on Education and Training (ICET 2017)*, Atlantis Press.

Bonwell, C.C. and Eison, J.A. (1991) 'Active learning: Creating excitement in the classroom,' *1991 ASHE-ERIC Higher Education Reports*, ERIC Clearinghouse on Higher Education, George Washington University.

Bransford, J., Sherwood, R., Vye, N. and Rieser, J. (1986) 'Teaching thinking and problem solving: Research foundations,' *American Psychologist*, 41(10): 1078.

DiCarlo, S.E. (2009) 'Too much content, not enough thinking, and too little FUN!,' *Advances in Physiology Education*, 33(4): 257–264.

Eberlein, T., Kampmeier, J., Minderhout, V., Moog, R.S., Platt, T., Varma-Nelson, P. and White, H.B. (2008) 'Pedagogies of engagement in science,' *Biochemistry and Molecular Biology Education*, 36(4): 262–273.

Essop, M.F. (2020) 'Implementation of an authentic learning exercise in a postgrad-uate physiology classroom setting,' *Advances in Physiology Education*, 44(3): 496–500.

Essop, M.F. and Beselaar, L. (2020) 'Student response to a cooperative learning element within a large physiology class setting: Lessons learned,' *Advances in Physiology Education*, 44(3): 269–275.

Essop, M.F. (2019) 'Autonomy pathways to compare active teaching methods in undergraduate physiology classes,' *Teaching and Learning Seminars*. Available at http://www.sun.ac.za/english/learning-teaching/ctl/t-l-resources/t-l-seminars (accessed 27 May 2021).

Foshay, R. and Kirkley, J. (1998) 'Principles for teaching problem solving. Technical paper #4,' *PLATO, TRO Learning*, Inc. Available at https://files.eric.ed.gov/fulltext/ED464604.pdf (accessed 27 May 2021)

Goodman, B.E., Barker, M.K. and Cooke, J.E. (2018) 'Best practices in active and student-centered learning in physiology classes,' *Advances in Physiology Education*, 42(3): 417–423.

Hmelo-Silver, C.E., Kapur, M. and Hamstra, M. (2018) 'Learning through problem solving,' In *International Handbook of the Learning Sciences*, Routledge, pp. 210–220.

Jonassen, D.H. (1997) 'Instructional design models for well-structured and Ill-structured problem-solving learning outcomes,' *Educational Technology Research and Development*, 45(1): 65–94.

Kamran, A., Rehman, R. and Iqbal, A. (2011) 'Importance of clinically oriented problem solving tutorials (COPST) in teaching of physiology,' *Rawal Medical Journal*, 36(3): 232–236.

Klegeris, A. and Hurren, H. (2011) 'Impact of problem-based learning in a large classroom setting: Student perception and problem-solving skills,' *Advances in Physiology Education*, 35(4): 408–415.

Lemoine, M. and Pradeau T. (2018) 'Dissecting the meanings of "physiology" to assess the vitality of the discipline,' *Physiology*, 33: 236–245.

Maton, K. (2014) *Knowledge and knowers: Towards a realist sociology of education*, Routledge.

Maton, K. (2016) 'Starting points: Resources and architectural glossary,' in K. Maton, S. Hood, and S. Shay (eds) *Knowledge building: Educational studies in Legitimation Code Theory*, Routledge.

Maton, K. and Chen, R.T. (2016) 'LCT in qualitative research: Creating a translation device for studying constructivist pedagogy,' in K. Maton, S. Hood, and S. Shay (eds), *Knowledge building: Educational studies in Legitimation Code Theory*, Routledge.

Maton, K. and Howard S. K. (2018) 'Taking autonomy tours: A key to integrative knowledge-building,' *LCT Centre Occasional Paper* 1: 1–35.

Maton, K. and Howard, S. K. (2021a) 'Targeting science: Successfully integrating mathematics into science teaching,' in Maton, K., Martin, J. R. and Doran, Y. J. (eds) *Teaching Science: Knowledge, language, pedagogy*, Routledge, 23–48.

Maton, K. and Howard, S. K. (2021b) Animating science: Activating the affordances of multimedia in teaching, in Maton, K., Martin, J. R. and Doran, Y. J. (eds) *Teaching Science: Knowledge, language, pedagogy*, Routledge, 76–102.

Michael, J. (2007) 'What makes physiology hard for students to learn? Results of a faculty survey,' *Advances in Physiology Education*, 31(1): 34–40.

Michael, J. and McFarland, J. (2011) 'The core principles ("big ideas") of physiology: Results of faculty surveys,' *Advances in Physiology Education*, 35(4): 336–341.

Michel, N., Cater III, J.J. and Varela, O. (2009) 'Active versus passive teaching styles: An empirical study of student learning outcomes,' *Human Resource Development Quarterly*, 20(4): 397–418.

NCCSTS (n.d.) National Center for Case Study Teaching in Science, Available at http://sciencecases.lib.buffalo.edu/ (Accessed 5 May 2021)

Priemer, B., Eilerts, K., Filler, A., Pinkwart, N., Rösken-Winter, B., Tiemann, R. and Zu Belzen, A.U. (2020) 'A framework to foster problem-solving in STEM and computing education,' *Research in Science and Technological Education*, 38(1): 105–130.

Rehan, R., Ahmed, K., Khan, H. and Rehman, R. (2016) 'A way forward for teaching and learning of Physiology: Students' perception of the effectiveness of teaching methodologies,' *Pakistan Journal of Medical Sciences*, 32(6): 1468.

Rootman-le Grange, I. and Blackie, M.A.L. (2018) 'Assessing assessment: In pursuit of meaningful learning,' *Chemistry Education Research and Practice*, 19(2): 484–490.

Rothman, D. (2014) 'A tsunami of learners called Generation Z,' *Public Safety: A State of Mind*, 1(1): 1–5.

Science Cases, (n.d.) 'A can of bull? Do energy drinks really provide a source of energy?' Available at https://sciencecases.lib.buffalo.edu/files/energy_drinks.pdf (Accessed 27 May 2021).

Smith, K.A., Sheppard, S.D., Johnson, D.W. and Johnson, R.T. (2005) 'Pedagogies of engagement,' *Journal of Engineering Education*, 94(1): 87–101.

Snyder, L.G. and Snyder, M.J. (2008) 'Teaching critical thinking and problem solving skills,' *The Journal of Research in Business Education*, 50(2): 90.

Wingfield, S.S. and Black, G.S. (2005) 'Active versus passive course designs: The impact on student outcomes,' *Journal of Education for Business*, 81(2): 119–123.

Part IV
Mathematical Sciences

10 A conceptual tool for understanding the complexities of mathematical proficiency

Ingrid Rewitzky

Introduction

In Mathematics modules at my institution, we challenge our students not only to know what certain definitions and results hold about a mathematical object but also to know why and to know how to represent and reason mathematically. Our students may convey their personal knowledge of a mathematical object by a picture or an example of their own reasoning process. Through effective teaching and engagement with the rhetoric of Mathematics, these personal representations and intuitive reasoning processes may be shaped into mathematical representations and mathematical reasoning (Ernest 1999). However, many students may resist such 'shaping' with questions such as: 'Where will this mathematical object or result be used in my studies and in real life?' This may seem to be about a particular object, result or topic in a particular branch of Mathematics. For many students, it is the level of abstraction that raises the question and affects mathematical proficiency. When I respond, I rarely provide an actual application but rather illustrate that a particular branch of Mathematics is useful as a whole. For example, calculus contributes to many aspects of daily life, including the design of bridges, the modelling behind weather forecasts or the computer program that determines investment portfolios. However, despite this real-world presence, calculus is also invisible in the sense that the underlying mathematical calculations are not actually seen. Another example is that electronic circuit design found in calculators and mobile phones makes extensive use of so-called imaginary numbers. The informal definition of imaginary numbers is that they are constructed on the basis of something that seems impossible – namely, the square root of minus one. Nevertheless, imaginary numbers have significant uses.

Over the years there have been many different opinions about how to represent all (not only some) mathematical objects. 'Mathematics is the domain within which we find the largest range of semiotic representation systems, both those common to any kind of thinking such as natural language and those specific to Mathematics such as algebraic and formal notations' (Duval 2006: 104). Therefore, Mathematics may be described as

DOI: 10.4324/9781003055549-14

having what Bernstein termed a 'horizontal knowledge structure' with 'strong grammars' – that is, 'a series of specialized languages with specialized modes of interrogation and criteria for the construction and circulation of texts' (2000: 160). The 'horizontal knowledge structure' implies independence of these specialized languages which include probability, algebra, logic, graphs. Each language is unique and does not displace or disprove another language. Moreover, each language has an 'explicit conceptual syntax' (Bernstein 2000: 163) and strong and recognizable principles. This makes it challenging to move between different specialized languages and from informal to formal representations in a specialized language. Mathematical reasoning is how we discover, formulate, justify and generalize claims about mathematical objects. The reasoning techniques are intrinsically linked to the mathematical representation used for referring to the mathematical objects. The complexity of the reasoning may be algorithmic in the case of illustrating that a claim holds for a particular instance of the mathematical object or a formal proof in the case of justifying that a claim holds for the mathematical object.

Throughout its history, Mathematics has been analyzed by philosophers, mathematicians and educationalists in order to better understand the nature of Mathematics and the difficulties experienced by many students in developing a deep and intuitive understanding of Mathematics.

Devlin has argued that we all possess the ability to cope with Mathematics provided we recognize what is required:

> To my mind, a limitation in coping with abstraction presents the greatest barrier to doing mathematics. And yet, as I shall show, the human brain acquired this ability when it acquired language, which everyone has. Thus the reason most people have trouble with mathematics is not that they don't have the ability but that they cannot apply it to mathematical abstractions.
>
> (Devlin 2000: 11)

Abstraction may involve starting with an entity or activity in our reality or world, abstracting the essential idea, features, and structure, understanding these as deeply and completely as possible and then defining a mathematical object and mathematical operations.[1] However, not all Mathematics derives its definition from representations of physicality. Central to Mathematics are generalizations or different levels of abstraction which involve moving from abstract representations to more encompassing abstract representations. Mason (1996) has argued that the ability to generalize is intrinsic to our success in Mathematics because it enhances our capability to apply mathematical concepts across mathematical tasks.

Duval (2006) has described developing mathematical proficiency as the ability (a) to distinguish a mathematical object from its representation, even though the only way that the mathematical object may be accessed is through its representation, and (b) to move between different mathematical representations of the mathematical object. Mathematical proficiency has been defined as having five interdependent components:

(a) *conceptual understanding* – comprehension of mathematical concepts, operations, and relations (b) *procedural fluency* – skill in carrying out procedures flexibly, accurately, efficiently, and appropriately (c) *strategic competence* – ability to formulate, represent, and solve mathematical problems (d) *adaptive reasoning* – capacity for logical thought, reflection, explanation, and justification, and (e) *productive disposition* – habitual inclination to see mathematics as sensible, useful, and worthwhile, coupled with a belief in diligence and one's own efficacy.

(Kilpatrick *et al.* 2001: 116)

Another description of mathematical proficiency is given in terms of 'what someone knows, can do, and is disposed to do mathematically' (Schoenfeld and Kilpatrick 2008: 326). Furthermore, there is the caution that assessment of mathematical knowledge may be complex, but assessment of strategic competence, adaptive reasoning and the ability to make mathematical connections may be even more difficult (Schoenfeld 2007).

There have been numerous studies on student and societal perceptions of the nature and value of Mathematics and the effect on developing mathematical proficiency. See, for example (MacBean 2004; Wood *et al.* 2012) and other related research described in those papers. MacBean notes

Many factors affect the quality of student learning. The students' conceptions of and approaches to learning, their prior experiences, perceptions and understanding of their subject, and the teaching and learning context can all influence the learning outcomes achieved.

(MacBean 2004: 553)

MacBean suggests that 'the more students believe that mathematics is integrated and integral to their degree course the more motivated they are likely to be, and the more meaning oriented their approaches to studying will become' (MacBean 2004: 562). Furthermore, Wood suggests that students be encouraged to

... appreciate the mathematics of all cultures and the contribution of mathematical ideas to the 'business of making accessible the richness of the world we are in, of making dense and substantial our ordinary, day-to-day living in a place – the real work of culture.'

(Wood 2000: 4)

In this chapter, I will illuminate four different insights of mathematical proficiency in terms of *what* mathematical knowledge and *how* one thinks/

reasons mathematically. For the what mathematical knowledge I distinguish between *knowing that* vs *knowing why* and for the *how* one thinks/reasons mathematically I distinguish between thinking/reasoning within Mathematics vs beyond Mathematics. Legitimation Code Theory (LCT) provides conceptual tools to understand these complexities of mathematical proficiency and for a differentiated support model for the sustained improvement of the learning experience in Mathematics.

Methodology

My starting point is the Specialization dimension of LCT, which highlights that every knowledge practice, belief or knowledge claim 'is about or oriented towards something and made by someone' (Maton 2014: 29). The organizing principles of these knowledge-knower structures can be conceptualized as specialization codes, generated by *epistemic relations* and *social relations*. For knowledge claims, epistemic relations are between knowledge and objects towards which the claim is oriented, and social relations are between knowledge and individuals conveying or making the claim. Each of these relations can be stronger or weaker along a continuum. For example, in developing mathematical proficiency, specialized knowledge is emphasized (stronger epistemic relations) while personal experience and opinions of students or lecturers are downplayed (weaker social relations) since not all mathematical ideas can be related to a personal experience or opinion. The two continua together generate *specialization codes*, illustrated in Figure 10.1: *knowledge codes* emphasize specialized knowledge, *knower codes*

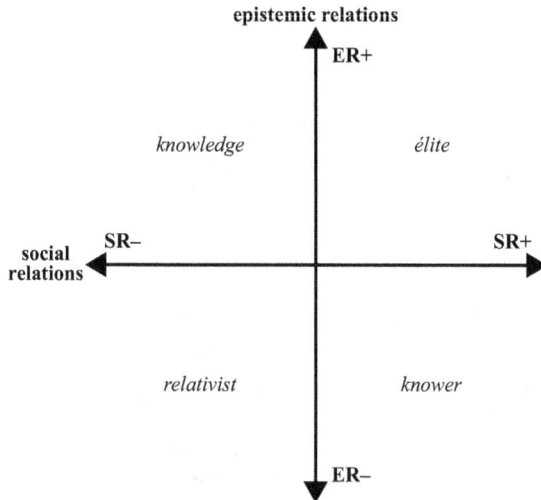

Figure 10.1 The specialization plane (Maton 2014: 30).

emphasize the right kind of knower, *élite codes* emphasize both specialized knowledge and the right kind of knower and *relativist codes* are 'anything goes' (Maton 2014).

The Specialization dimension has provided insights into, for example, degrees of clash between: students' dispositions and educators' pedagogic practices, different approaches within a specific discipline, curriculum and pedagogy of a discipline (Maton 2014, Maton and Chen 2020).

The epistemic relations focus specifically on the nature of knowledge and provide a means to consider the relationship between the *what* and *how* of a knowledge practice. This relationship, explored by the *epistemic plane*, differentiates between *ontic relations* (OR) that represent the strength of relations between a knowledge claim and the object of study and *discursive relations* (DR) that represent the strength of relations between different ways of referring to or dealing with objects of study (Maton 2014). Each relation can be independently stronger (+) or weaker (–) along a continuum. When brought together, the two strengths generate four *insights*. For *purist insights*, practice is based on strong adherence to both a strongly distinguished object of study (OR+) and strongly distinguished approach (DR+). For *doctrinal insights*, practice is not governed by a distinctive object of study (OR–) but by a strongly differentiated approach (DR+). For *situational insights*, knowledge practices are specialized by a distinctive object of study (OR+) and by relative freedom as to how this object is studied (DR–). For *knower/no insights*, practice is either characterized by 'anything goes' (neither a differentiated object nor a differentiated approach; OR–, DR–) or, where these weaker epistemic relations are paired with stronger social relations, legitimated through attributes of the knower.

This *epistemic plane*, together with reflections on my own experience as a mathematician, inspired the theoretical framework I shall use in this chapter. I adapt ontic relations (OR) to refer to '*what* mathematical knowledge' and discursive relations (DR) to refer to '*how* one thinks/reasons.' (This is only one way these concepts can be used and reflect my concerns in this chapter.) It is expected that for a mathematical object of study in a Mathematics module, students' knowledge of a certain mathematical claim about the mathematical object may vary from not knowing that the claim holds to simply knowing that the claim holds to knowing why the claim holds. Also, a student's way of expressing their understanding may have stronger or weaker levels of mathematical formalism. The stronger and weaker ontic relations and discursive relations may be identified along a continuum of mathematical knowledge and a continuum of more or less mathematical formalism, respectively, as in Table 10.1. At right angles to each other, these continua form four quadrants each representing an insight of mathematical proficiency, as previously depicted in Figure 10.2.

Table 10.1 Ontic relations and discursive relations for a particular mathematical
object of study

Ontic relations (OR)	Strength of relations between a knowledge claim and the mathe-matical object of study	+ *stronger*	Know why a mathematical claim holds for the mathematical object of study
		– *weaker*	Know that a mathematical claim holds for the mathematical object of study
		-- weakest	Do not know that a mathematical claim holds for the mathematical object of study
Discursive relations (DR)	Strength of relations between different ways of referring to or reasoning about the mathematical object of study	+ *stronger*	Think/reason with examples, representations, and techniques from mathematics
		– *weaker*	Think/reason with examples, representations, and techniques from beyond mathematics

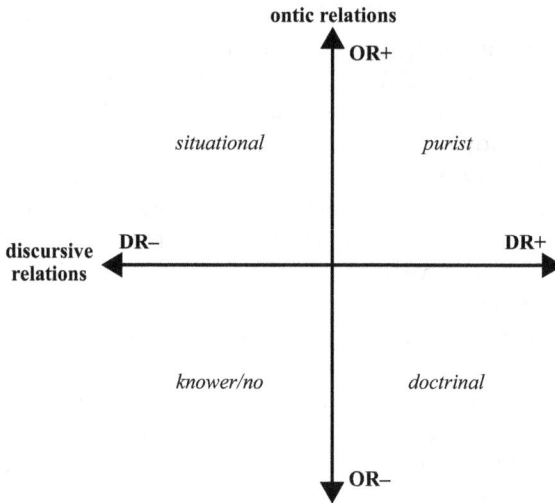

Figure 10.2 The epistemic plane (Maton 2014: 177).

There is no right or wrong *insight*. However, when embarking on a math-ematical study, a certain insight may be a preferred starting point over the others. Success in mathematical thinking and reasoning requires 'insight shifting' which involves strengthening or weakening ontic relations and/or discursive relations. In particular,

- *abstract* from personal representation and intuitive reasoning into more mathematical representations and formal reasoning techniques (towards stronger discursive relations);
- *acquire* knowledge of the underlying mathematical ideas and principles (towards stronger ontic relations);
- *generalize* from a specific instance of a mathematical concept or technique to a more encompassing mathematical concept or technique (move towards weaker ontic relations and stronger discursive relations, thereby shifting from situational insight to doctrinal insight);
- *specialize* from a general concept to instances of the concept (move towards weaker ontic relations and so from purist insight to situational insight); and
- *link or apply*, if possible, the mathematical knowledge and skills to a personal experience, perspective or a real-world phenomenon (move to knower/no insight).

Each of these shifts provides a significant challenge when developing mathematical proficiency in a mathematical topic. Different levels of Mathematics proficiency may be described in terms of the different insights navigated. In particular, a basic level of mathematical proficiency will be entirely in *doctrinal insight*; an intermediate level of mathematical proficiency will draw on two different insights – typically, *doctrinal insight* and *situational insight*; and a high level of mathematical proficiency will draw on three different insights as the needs for abstraction, generalization or specialization demand.

Without effective strategies for facilitating insight shifting, there is a potential for clashes. For example, mathematicians may be working in purist insight while most students may be entirely in doctrinal insight or situational insight where less mathematical formalism and rigour are used. Another potential clash arises for mathematical topics that have emerged entirely in purist insight without any link to experiential phenomena.

Analysis of Mathematics proficiency using the epistemic plane

Four case studies have been selected, based on what is studied in a typically undergraduate Mathematics module, namely, mathematical objects, mathematical activities, mathematical representations and mathematical structures. Each will be briefly introduced and an example will be analyzed with the emphasis on the shifting between insights needed for developing mathematical proficiency and the challenges that are typically experienced.

A mathematical object

A mathematical object may be an abstraction of a real-world object or a generalization of an existing mathematical object. As an example, consider the mathematical object called a *function*. At first, a 'function' (a term due to Leibnitz (1646–1716)) simply meant a dependence of numbers given by an

analytic expression (situational insight). As the need for formalism and generality grew, the notion of a function went through a gradual transition to the
abstract notion of a function between two sets, which was first introduced by
Richard Dedekind (1831–1916) (doctrinal insight). However, when setting
up a family tree relationship of family members, ancestors and relatives, it will
be observed, for example, that the 'mother-of' relationship may relate one
family member to more than one other family member (knower/no insight).
Such relationships are captured by the abstract notion of binary relation,
introduced by Augustus de Morgan (1860) and Peirce's logic of relatives
(1870), which generalizes the notion of a function (purist insight). The biggest challenge here is understanding that the purist insight of a function and
the doctrinal insight of a function are equivalent. A strategy for developing
mathematical proficiency in a mathematical object (in general) is depicted in
Figure 10.3 and may involve the following sequence: situational insight to
doctrinal insight to knower/no insight to purist insight to doctrinal insight.

A mathematical activity

A mathematical activity typically originates from a practical experience such as,
for example, measuring or counting. Let us consider the mathematical activity
of finding areas. All first-year Mathematics students will know that the area of
a circle may be found by using the formula $A = \pi r^2$ and also know how to use
that formula mathematically, for example, to find the area of a circle given its
radius or its diameter (doctrinal insight). Unfortunately, few students will
know why this is the case. However, people have known, at least since biblical
times, that there is a way to divide a cake or a piece of land between two people so that neither is envious of the other – one person cuts and the other
chooses (knower/no insight). Properties of the whole may be described in
terms of properties of the two parts. Abstracting from this to a circle cut into

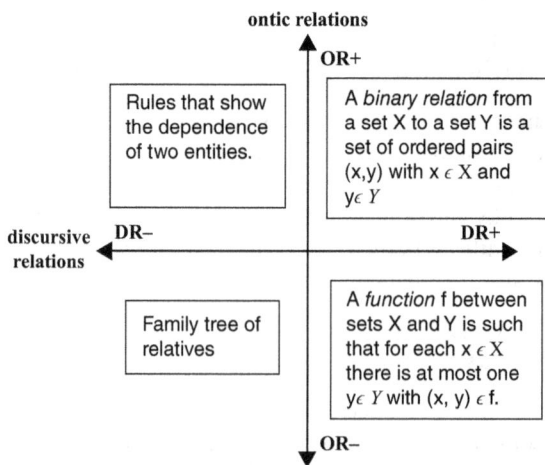

ontic relations

OR+

| Rules that show the dependence of two entities. | A *binary relation* from a set X to a set Y is a set of ordered pairs (x,y) with x ϵ X and yϵ Y |

discursive relations   DR–   DR+

| Family tree of relatives | A *function* f between sets X and Y is such that for each x ϵ X there is at most one yϵ Y with (x, y) ϵ f. |

OR–

Figure 10.3 Insights for functions.

segments, the area of the circle may be found as the sum of the areas of those segments (situational insight). Reasoning formally in terms of the Riemann sum and the definition of a definite integral yields a formal derivation of the formula (purist insight). Therefore, a strategy for developing mathematical proficiency in a mathematical activity (in general) is depicted in Figure 10.4 and may involve the following sequence: doctrinal insight to knower/no insight to situational insight to purist insight. The biggest challenge is the generalization needed to move from situational insight to purist insight.

Linking two mathematical representations of a mathematical object

A mathematical object may have different mathematical representations, depending on the perspective or branch of Mathematics in which it is being explored. The mathematical object called a torus arises in different branches of Mathematics including calculus and topology and also beyond Mathematics in astrophysics, biology medicine, nuclear physics. Mathematical proficiency in each mathematical representation would be needed before being able to link them. A student may (a) have heard the expression that a coffee cup and doughnut are the same to a mathematician because they both have a single hole (knower/no insight), (b) investigate various torus-shaped objects (situational insight), (c) explore the basic calculus representation of a torus as the surface of revolution generated by revolving a circle in a three-dimensional space (doctrinal insight) and (d) move to a more sophisticated topological representation of a torus as a closed surface with a hole, defined by the product of two circles (purist insight). Therefore, a strategy for developing mathematical proficiency in linking mathematical representations of a mathematical object (in general) is depicted in Figure 10.5 and may involve the following sequence: knower/no insight to situational insight to doctrinal insight to purist insight.

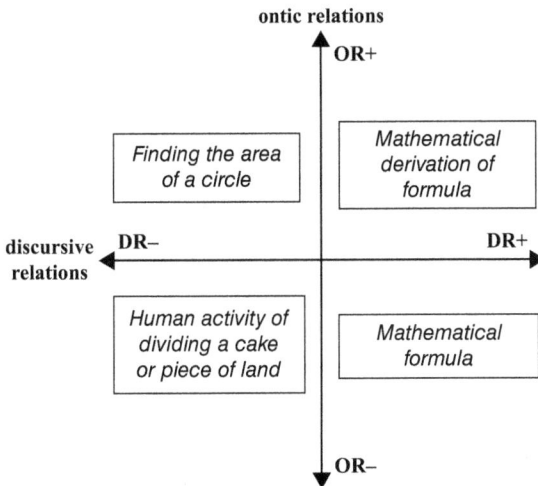

Figure 10.4 Insights for a mathematical activity.

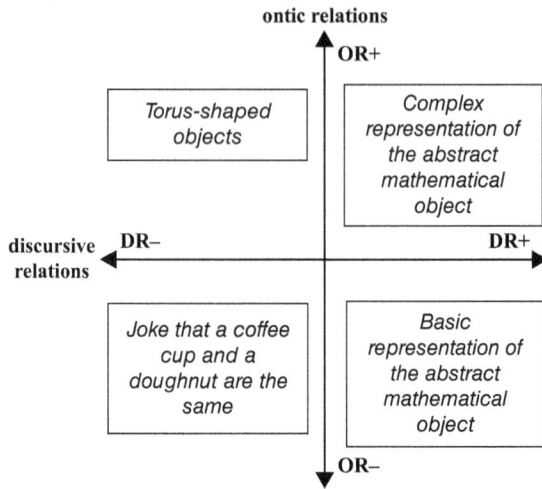

Figure 10.5 Insights for linking mathematical representations of a mathematical object.

The biggest challenges typically arise in generalizing from situational insight to doctrinal insight or in abstracting from doctrinal insight to purist insight.

A mathematical structure

Mathematical structures express mathematical principles or abstractions intended to capture generic properties about a collection of objects (situational insight). A mathematical structure may be likened to a human skeleton. The skeleton is the basic structure of the human body. Although the outward appearances of people may differ, the inward structure, the shape and arrangement of the bones are the same (knower/no insight). Similarly, mathematical structures represent the underlying sameness in situations that may appear outwardly different. Following the Hilbert programme of 1920 and assuming set theory, a mathematical structure is a formal axiomatic system consisting of vocabulary of symbols and connectives, axioms capturing properties of certain symbols and connectives, and rules for combining symbols and connectives and reasoning about them (purist insight). For example, the collection of real numbers with designated symbols 0 and 1, operations of addition and multiplication and axioms of associativity, commutativity, identity is a familiar mathematical structure (doctrinal insight). Therefore, a strategy for developing mathematical proficiency in mathematical structures, depicted in Figure 10.6, may involve linking an informal and a formal yet familiar mathematical structure and then linking the formal yet familiar mathematical theory with formal more abstract mathematical structure.

These strategies may support students to develop mathematical proficiency that has breadth and depth. It will have breadth in the sense of linking different mathematical representations and the associated reasoning techniques

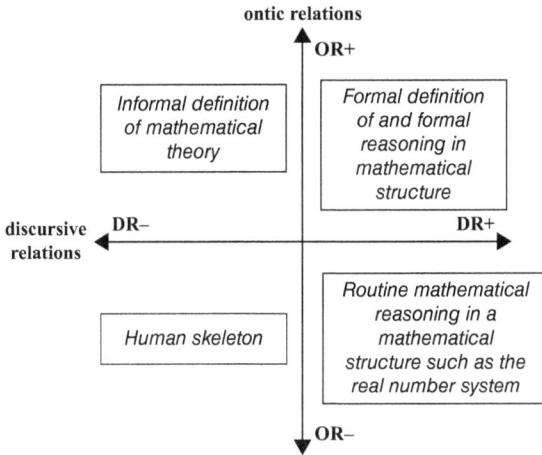

Figure 10.6 Insights for concept of mathematical structure.

at the same level of formalism and of transferring acquired mathematical knowledge and skills within or beyond Mathematics. It will have depth in the sense of understanding and using different levels of mathematical formalism and reasoning techniques.

Discussion and conclusion

Understanding the complexities of Mathematics proficiency of students entering tertiary education has been identified as important for improving student success in STEM programmes (Bohlmann *et al.* 2017; Council on Higher Education 2013; Scott *et al.* 2007). A mathematically proficient student typically has a productive disposition towards Mathematics – that is, the tendency to see sense in Mathematics, to perceive Mathematics as both useful and worthwhile, to believe that steady effort in learning Mathematics pays off and to be an effective doer of Mathematics (Kilpatrick *et al.* 2001).

Using the Specialization dimension of LCT, this chapter illuminates different insights of mathematical proficiency and possible strategies for developing mathematical proficiency necessary for students to be successful in STEM programmes. The key elements studied in Mathematics modules were analyzed in terms of *what* mathematical knowledge and *how* one thinks/ reasons, and the emerging insights were depicted on the epistemic plane.

Four key observations may be made. Firstly, moving horizontally, from weaker to stronger discursive relations (understood here as *how* one thinks/ reasons), corresponds to the challenge experienced with the nature of abstraction in Mathematics. Secondly, moving vertically, from weaker to stronger ontic relations (understood here as *what* mathematical knowledge), corresponds to the challenge experienced with the level of abstraction in Mathematics. Thirdly, mathematical proficiency presents differently in individual students and corresponds to the ability to shift between different insights. Fourthly, there are potential clashes when the preferred insight of the lecturer and the students differ.

The evidence of challenges with insight shifts and clashes suggests that the Mathematics curriculum could benefit from an analysis of the different insights of the mathematical objects, mathematical activities, mathematical representations and mathematical theories covered and from making these different insights explicit, especially where the biggest challenges are anticipated.

An approach for integrating into the curriculum the insight shifting strategies proposed in this chapter is what may be called a 'differentiated support' model. This model acknowledges different levels of mathematical proficiency of students and offers different levels of support and enrichment so that students have effective learning opportunities to reach their level of success along the path that best suits their style of learning. A key feature of the model is facilitating the navigation between the insights. In particular, for each Mathematics topic, there is a worksheet which is divided into three (or ideally four) parts: procedure problems for checking understanding of core concepts and reasoning for routine problems (doctrinal insight); principle questions for understanding the more theoretical and abstract components of the topic, for developing mathematical writing ability and for thinking critically about results (purist insight); a possibilities section with specific instances or applications to explore, alternative ways to explore a concept and a project (situational insight); and, if possible, reading material of applications in other disciplines or in the real world (knower/no insight). Students are supported and incentivized to submit or present their representation and reasoning approaches to the lecturer or tutor for feedback and there are multiple opportunities for fine-tuning.

This differentiated support model may also provide a valuable framework to guide the planning of technology-mediated support initiatives. Based on the profile of a student, a personalized suite of compulsory and optional learning opportunities for development and growth in mathematical proficiency could be offered throughout the year. It could give students the freedom to choose what they would like to do and when, and develop mathematical proficiency in areas identified through assessments for determining proficiency gaps as well as those that have been self-identified and which are of particular interest to the student.

It is hoped that the insights of this chapter will contribute to improving mathematical proficiency and will be of value to other fundamental disciplines in Science.

Note

1 For example, from the human activity of counting, the mathematical objects called numbers are abstracted together with algebraic operations. Mathematical objects are abstract, unobservable and on the platonic view exist independently. Mathematical representations are the way we refer to mathematical objects. If we think about how to deal with mathematical objects (such as numbers, functions, relations, fields) it is not possible to perceive, manipulate or work with a mathematical object without its mathematical representations. For example, we cannot 'see' a function without its algebraic expression or its graph.

References

Bernstein, B. (2000) *Pedagogy, symbolic control and identity: Theory, research, critique*, Rowman and Littlefield.

Bohlmann, C.A., Prince, R.N. and Deacon, A. (2017) 'Mathematical errors made by high performing candidates writing the national benchmark tests', *Pythagorus – Journal of the Association for Mathematics Education in South Africa*, 38(1): 1–10.

Council on Higher Education. (2013) *A proposal for undergraduate curriculum reform in South Africa: The case for a flexible curriculum structure*, Council on Higher Education.

Devlin, K. (2000) *The math gene: How mathematical thinking evolved and why numbers are like gossip*, Basic Books.

Duval, R. (2006) 'A cognitive analysis of problems of comprehension in a learning of mathematics', *Educational Studies in Mathematics*, 62: 103–131.

Ernest, P. (1999) 'Forms of knowledge in mathematics and mathematics education: Philosophical and rhetorical perspectives', *Educational Studies in Mathematics*, 38: 67–83.

Kilpatrick, J., Swafford, J and Findell, B. (2001) 'The strands of mathematical proficiency', in J. Kilpatrick, J. Swafford, and B. Findell (eds) *Adding it up: Helping children learn mathematics*, National Academy Press.

MacBean, J. (2004) 'Students' conceptions of, and approaches to, studying mathematics as a service subject at undergraduate level', *International Journal of Mathematical Education in Science and Technology*, 35(4): 553–564.

Mason, J. (1996) 'Expressing generality and roots of algebra', in N. Bednarz, C. Kieran, and L. Lee (eds) *Approaches to algebra: Perspectives for research and teaching*, Kluwer Academic Publishers.

Maton, K. (2014) *Knowledge and knowers: Towards a realist sociology of education*, Routledge.

Maton, K. and Chen, R.T-H. (2020) 'Specialization codes: Knowledge, knowers and student success', in J.R. Martin, K. Maton, and Y.J. Doran (eds) *Accessing Academic Discourse: Systemic Functional Linguistics and Legitimation Code Theory*, Routledge.

Scott, I., Yeld, N. and Hendry, J. (2007) A case for improving teaching and learning in South African higher education, Higher Education Monitor No. 6: The Council on Higher Education.

Schoenfeld, A. H. (2007) 'Reflections on an assessment interview: What a close look at student understanding can reveal', in A. H. Schoenfeld (ed) *Assessing Mathematical Proficiency*, Cambridge University Press.

Schoenfeld, A.H. and Kilpatrick, J. (2008) 'Toward a theory of proficiency in teaching mathematics', in T. Wood and D. Tirosh (eds) *International Handbook of Mathematics Teacher Education Volume 2: Tools and Processes in Mathematics Teacher Education*, Sense Publishers.

Wood, L.N. (2000) 'Communicating mathematics across cultures and time', in H. Selin (ed) *Mathematics across Cultures: The History of Non-Western Mathematics*, Kluwer Academic Publishers.

Wood, L.N., Mather, G., Petocz, P., Reid, A., Engelbrecht, J., Harding, A., Houston, K., Smith, G.H. and Perrett, G. (2012) 'University students' views of the role of mathematics in their future', *International Journal of Science and Mathematics Education*, 10(1): 99–119.

11 Supporting the transition from first to second-year Mathematics using Legitimation Code Theory

Honjiswa Conana, Deon Solomons, and Delia Marshall

Introduction

The transition from first to second year is identified as a challenge for many students in undergraduate programmes around the world (see, for example, Hunter *et al.* 2010 in the USA context; Yorke 2015 in the UK context). Yet, there is surprisingly little research on this transition, with most studies focusing either on students' experiences of the first year or on students' exit-level outcomes. In South Africa, likewise, concerns about transition tend to be concentrated at the transition from school to first-year university. However, studies have argued that there are key 'epistemic transitions' (Council on Higher Education (CHE) 2013) throughout the undergraduate degree that need pedagogical attention.

In this chapter, we illustrate how concepts from Legitimation Code Theory (LCT; Maton 2014) were used to develop insights into the challenges that STEM students face in making the transition to second year. These insights then were used to frame an educational intervention in second-year Mathematics, and this chapter reports on the impact this intervention had on students' learning.

Context of the study

This study took place in the Faculty of Natural Sciences at an historically black South African university. Many of the students are first-generation students in higher education, meaning that their parents/guardians had not attended a tertiary institution. About a third of the first-year intake of BSc students is placed in an extended curriculum programme (ECP). This ECP is a four-year BSc degree, which essentially enables students to complete the first year of their BSc degree over two years, with foundation provision embedded in Physics and Mathematics courses (including strengthening conceptual understanding, strengthening academic literacy in engaging with science texts, etc.). Despite the extensive foundational provision of the ECP, which aimed to give students a solid foundation in Physics and Mathematics, students' transition to second-year Physics and Mathematics remained an ongoing challenge. Student motivation appeared to decline in second year,

DOI: 10.4324/9781003055549-15

accompanied by poor pass rates; this was the case both for ECP students and for those entering second year via the regular three-year BSc route. One obvious reason for the students' transition challenges is that the second year Physics and Mathematics courses become more mathematically demanding. In addition, students are required to apply the Mathematics learning from their Mathematics courses to their Physics courses, which many find challenging (Bing and Redish 2009).

In order to better understand students' transition challenges, first- and second-year Physics and Mathematics classes were observed and interviews conducted with second-year students. Tools from LCT were useful in characterizing the teaching practices in these courses and in beginning to identify some of the obstacles that students were experiencing in making the transition to second year.

Legitimation Code Theory as a tool for thinking about transition to second year

LCT is a sociological 'toolkit' (Maton 2014: 15) which integrates and extends key concepts from, among others, the work of sociologists Basil Bernstein and Pierre Bourdieu, including Bernstein's code theory, knowledge structures and pedagogic device (Bernstein 1996), and Bourdieu's concepts of field theory, capital and habitus (Bourdieu 1994); for a more detailed account of the develop of LCT, see Maton (2014). LCT comprises three active dimensions or sets of concepts that explore different organizing principles underlying practices, dispositions and context. For the purposes of this chapter, we focus on two dimensions: Semantics and Specialization. Concepts from the Semantics and Specialization dimensions provide useful tools to characterize STEM knowledge structures and practices, in order to begin to tease out some of the transition challenges that students experience. In this section, we provide a brief overview of the key concepts from the Semantics and Specialization dimensions used in our study.

Concepts from Semantics for thinking about transition to second year

Concepts from the Semantics dimension of LCT (Maton 2009, 2013, 2014, 2020) provide useful conceptual tools for allowing us to analyze the knowledge structure and practices of STEM disciplines. *Semantic gravity* is defined as the extent to which meaning 'is related to its context of acquisition or use' (Maton 2009: 46). When semantic gravity is weaker, meaning is less dependent on its context. In other words, semantic gravity is related to the degree of abstraction. For example, the decontextualized, abstract Physics concept of 'force' can be applied to a wide range of specific contexts, ranging from vast galaxies to tiny atoms.

Semantic density describes the complexity of meanings and is defined as the extent to which meaning is concentrated or condensed within symbols (a term, concept, phrase, expression, gesture, etc.) (Maton 2014). In STEM disciplines, meaning is often condensed within nominalizations (scientific words or phrases that are dense in meaning), such as 'acceleration' (Physics), 'photosynthesis' (Biology). A great deal of information is also condensed into graphs, symbols, diagrams and mathematical equations.

To visualize the relative strengths of semantic gravity (SG) and semantic density (SD) over time, Maton (2013, 2014, 2020) developed the analytical method of *semantic profiling*. This indicates in the form of a diagram how strengths of SG and SD vary over time. The strengths of SG and SD are represented on the *y*-axis, with time on the *x*-axis. In the semantic profile, SG and SD are typically portrayed as inversely related, though this may not always apply or be analytically appropriate. In such cases, either drawing separate profiles or representing SG and SD on a *semantic plane*, allowing SG and SD to vary independent of each other (see Maton 2014; Blackie 2014, for a Chemistry example). The *semantic profile* can be used to map practices as they unfold in time, whether in a student task (e.g. an essay or problem task), a single classroom episode, part of a lesson, a series of lessons, an entire course or even a whole curriculum. Figure 11.1 shows three different semantic profiles: if these corresponded to three different lessons, then A1 would represent a lesson in which the teaching remained at the level of general principles, representing weaker semantic gravity and stronger semantic density (SG−, SD+); A2 would represent a lesson that remained at the level of specific examples, representing stronger semantic gravity and weaker semantic density (SG+, SD−); B would indicate a lesson where there was shifting in semantic gravity (context-dependence) and semantic density (complexity) through unpacking and repacking of representations. Profile B is said to have a greater 'semantic range' than either A1 or A2.

Variations in strengthening and weakening of semantic gravity and semantic density, Maton (2013) argues, is one way that meaningful learning is enabled. Many teachers of science go from the level of a general principle down into specific examples but never connect the examples back to the

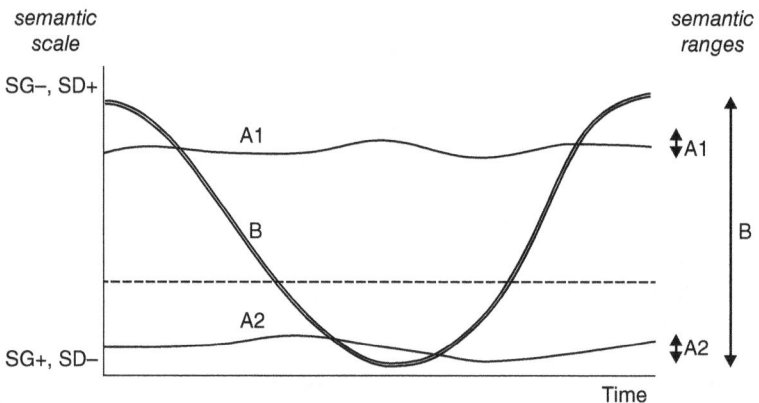

Figure 11.1 Illustrative profiles and semantic ranges (Maton 2013: 13).

underlying principle. Maton terms this a 'down escalator' profile because the teacher repeatedly 'unpacks' and simplifies technical concepts and relates these to specific examples, yet never models the process of shifting upward, through condensing meaning into technical terms or relating specific examples to the general principles (Maton 2013: 17). In this study, the method of semantic profiling is used to characterize the teaching practices in Physics and Mathematics courses.

Concepts from Specialization for thinking about transition to second year

In analyzing the form taken by knowledge in various disciplines, Bernstein (1996) introduced the concept of 'knowledge structure' and distinguished between 'hierarchical' and 'horizontal' knowledge structures. STEM disciplines (such as Physics, Chemistry, Biology) are typically characterized as 'hierarchical knowledge structures,' each being 'an explicit, coherent, systematically principled and hierarchical organization of knowledge' (Bernstein 1996: 172). Horizontal knowledge structures, on the other hand, are those which consist of 'a series of specialized languages, each with its own specialized modes of interrogation and specialized criteria' (Bernstein 1996: 172). Bernstein classifies Mathematics as a horizontal knowledge structure since 'it consists of a set of discrete languages for particular problems' (Bernstein 2000: 165); As Wolff elaborates, the 'languages' of Mathematics (for example, geometry, calculus, trigonometry, algebra) each have their own principles and procedures. They 'need to be acquired independently, and do not necessarily relate to each other or integrate concepts across the languages' (Wolff 2015: 39). Mathematics possesses what Bernstein terms a 'strong grammar,' meaning that its languages 'have an explicit conceptual syntax' (Bernstein 2000: 163). This is in contrast to horizontal knowledge structures with 'weak grammar' (for example, disciplines within the arts and humanities).

The Specialization dimension of LCT, with its concept of 'knowledge-knower structures,' usefully expands on Bernstein's conceptualization of knowledge structures by asserting that each knowledge structure also has an expectation, whether explicit or tacit, of a certain kind of ideal knower (Maton 2014). Specialization is based on the assumption that every social/educational practice is oriented towards something (knowledge) and by someone (knower). For each practice, it is necessary to identify '*what* can be legitimately described as knowledge (epistemic relations); and *who* can claim to be a legitimate knower (social relations)' (Maton 2014: 29). Epistemic relations (ER) and social relations (SR) in any practice can be stronger (+) or weaker (−), and can be represented on the *specialization plane,* as shown in Figure 11.2. By examining the epistemic relations and social relations of a particular practice, its position on the plane can be seen to fall into one of four quadrants – described as *knowledge codes, knower codes, élite codes* and *relativist codes.*

STEM disciplines are typically characterized as being characterized by stronger epistemic relations (ER+) since the scientific knowledge that the

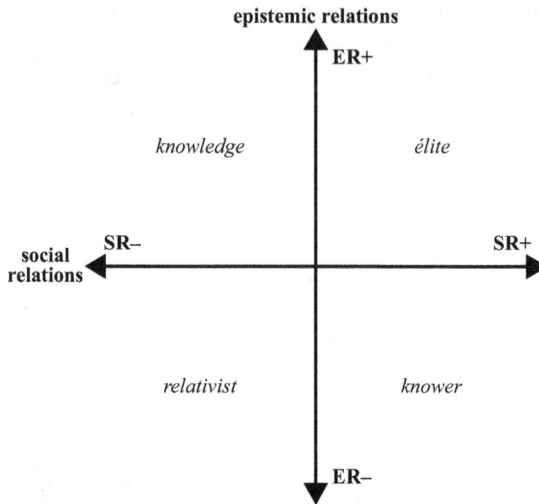

Figure 11.2 The specialization plane (Maton 2014: 30).

scientist possesses is emphasized and the attributes or dispositions of the scientist (or knower) are downplayed, representing weaker social relations (SR–): a *knowledge* code. Typically, Mathematics is typically characterized as a knowledge code. In contrast, in (for example) some humanities disciplines, specialized knowledge may be downplayed (ER–) and attributes of knowers may be emphasized (SR+): a *knower* code. Music as a high school subject is considered an *élite* code (Lamont and Maton 2008) since musical knowledge is valued (ER+), as well as having a musical 'feel' or disposition (SR+).

The characterization of Mathematics as a knowledge code (where dispositions of knowers are downplayed) is in contrast to the popular belief in that 'innate ability' matters in Mathematics learning. Contemporary Mathematics education research (e.g. Boaler 2016) notes the prevalence of the belief held by schoolchildren, their parents and teachers that 'some people are born with a "math brain" and some are not, and that high achievement is only available to some students.' This belief in 'innate ability' suggests relatively strong social relations and therefore suggests that Mathematics is often perceived as an élite code. This belief in 'innate ability' is also linked to the widespread phenomenon of 'Mathematics anxiety,' which is defined as anxiety about one's ability to do Mathematics. Mathematics anxiety correlates negatively with confidence and motivation (Ma 1999; Ashcraft 2002). This perception of the 'innateness' of Mathematics ability can also be traced to historical accounts of the development of Mathematics as a discipline. De Freitas and Sinclair (2014) note how the Cartesian mind-body divide is still dominant in Mathematics. They argue that this belief – that 'intuition' and 'innate mental talent' are key for success in Mathematics – can be alienating and play a gate-keeping role for the discipline.

In summary, contemporary research on Mathematics education as well as historical accounts of the development of Mathematics as a discipline suggests that a perception of 'innate ability' dominates, positioning these perceptions of Mathematics within an élite code on the Specialization plane. The implications of this is that Mathematics education needs to explicitly challenge these perceptions of 'innateness' – with not just a focus on the acquisition Mathematics knowledge, but also on developing the knower dispositions for the discipline (see, for example, Boaler (2016) on developing students' mathematical mindsets).

In the context of undergraduate science, echoing Maton's emphasis that 'there are always knowledges and always knowers' (2014: 96), Ellery (2018, 2019) has challenged the notion that the emphasis on specialized knowledge should eclipse issues about knowers dimensions. Ellery distinguishes between *production-context* knowers (as a scientist) and *learning-context* knowers (as a science learner). Production-context knowers need to value the epistemic norms and values of science, including rigour, curiosity, objectivity, working accurately, thinking analytically and critically (Ellery 2018: 31). Learning-context knowers need to develop knower attributes appropriate for learning university science. These include dispositions such as working independently, reflecting on one's learning (being metacognitive) and adopting appropriate approaches to learning (Ellery 2018); in other words, focusing on deep approaches (developing conceptual understanding) rather than surface approaches (focusing on rote learning) (Marton and Säljö 1976).

Ellery argues that in many traditionally content-dominated STEM courses (with a strong knowledge code), there is not enough explicit focus on developing knower dispositions, values and ways of thinking important for success in the discipline: 'to become effective science learners, students need to acquire not only certain practices and knowledge (representing weaker epistemic relations) but also certain knower dispositions (representing stronger social relations)' (Ellery 2019: 231). Similarly, Mtombeni (2018) argues that the lack of focus on social relations in a first-year Chemistry curriculum limits the development of students' knower dispositions.

An LCT analysis of the transition to second year

In this section, we draw on concepts from the LCT dimensions of Semantics and Specialization to develop an understanding of the hurdles students face in making the transition to second year. Using the concepts of semantic gravity and semantic density, we present semantic profiles of some representative Physics and Mathematics lessons to highlight differences and discontinuities in teaching practices between first-year and second-year courses. We also draw on the Specialization concepts of knowledge-knower structures and specialization codes to illuminate some of the difficulties students face in succeeding in their second-year studies. Data for this section is drawn from a previous study (Conana *et al.* 2019), which constructed semantic profiles of lecture sequences in first-year ECP Physics and second-year Physics and

Mathematics and interviewed second-year students about their experiences in transition to second year.

Semantics analysis

In applying the concepts of semantic gravity and semantic density to the context of undergraduate Physics, we used what in LCT is termed a 'translation device' (Maton and Chen 2016) which helped to show how concepts are realized in the empirical data of the study. The translation device draws on the work of Lindstrøm (2012) and Georgiou (2014), who have presented ways of coding the relative strengths of semantic gravity in the context of Physics lectures and students' responses to Physics tasks. They use the label *abstract* to refer to statements of general principles or laws; *concrete* refers to a description of specific examples; intermediate (or linking) refers to instances where general principles and specific examples are linked. Table 11.1 presents the translation device for semantic gravity and semantic density used in this study to characterize pedagogic practices.

In constructing semantic profiles of lessons, data was drawn from classroom observation notes and video recordings of lectures. From this data, semantic profiles were constructed to map shifts in semantic gravity and semantic density during lessons, as lecturers moved between abstract principles and specific examples, as well as the ways in which representations were unpacked or condensed during each lecture. The relative strengths of SG and SD were characterized as concrete, linking or abstract. At the concrete level, the lecturer would be referring to specific examples (SG+) and representations would be unpacked, often in the form of a verbal representation (SD–). At the abstract level, the lecturer would be using new concepts or general principles (SG–), mostly represented in semantically denser modes (graphical, diagrammatic, mathematical). The linking level is characterized by the lecturer building on familiar concepts or principles in a linking way; in doing so, dense representations were being explicitly unpacked or repacked into their constituent parts or meaning. For details of the data reduction and analysis process, and the translation device used, see Conana *et al.* (2019).

Through mapping shifts in semantic gravity and semantic density, semantic profiles were constructed for several lessons. Here, we present three

Table 11.1 Translation device for various levels of Semantics

	Realizations of semantic gravity	*Category*	*Realizations of semantic density*	
SG– ↑ ↓ SG+	New general principles Familiar principles used in a linking way Specific examples	Abstract A` Linking (Intermediate) L Concrete C	Representations (or nominalizations) Unpacking or repacking representations Linking representations to specific examples	SD+ ↑ ↓ SD–

semantic profiles: one for an ECP Physics lesson (Figure 11.3), one for a second-year Physics lesson (Figure 11.4) and one for a second-year Mathematics lesson (Figure 11.5). On the semantic profiles, coding (in the form of line thickness) is used to indicate the different forms of interaction in lectures (with a thin line indicating where only the lecturer is talking and a thick line indicating lecturer-student interactions).

A more detailed analysis of these semantic profiles is given in Conana *et al.* (2019). The comparison of the first year and second year semantic profiles, together with data from student interviews, highlighted several key differences or discontinuities in teaching practices:

Firstly, the *semantic range* in the lessons diminishes with the transition to second year. As evident in Figure 11.3, the first year ECP teaching spans a

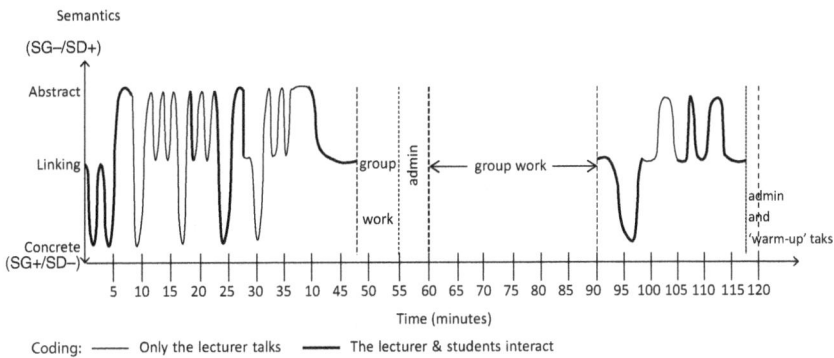

Figure 11.3 Semantic profile of a first-year ECP physics lesson.

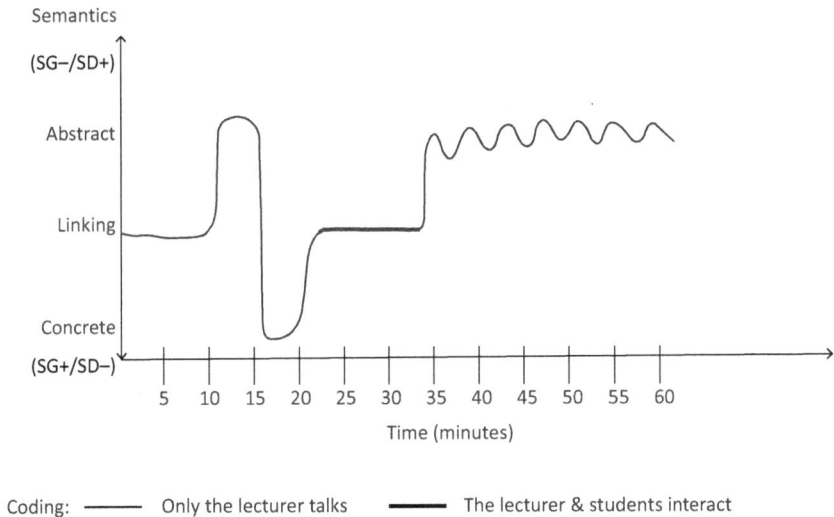

Figure 11.4 Semantic profile of a second-year physics lesson.

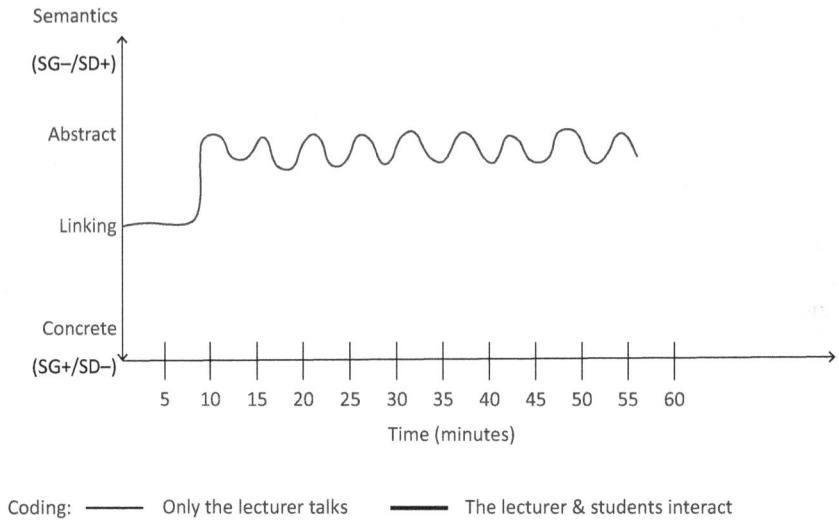

Coding: ———— Only the lecturer talks   ▬▬▬▬ The lecturer & students interact

Figure 11.5 Semantic profile of a second-year mathematics lesson.

large semantic range (spanning abstract and concrete). By contrast, the semantic range of the second-year Physics and Mathematics course is narrower, predominantly at the abstract level, which is to be expected in these more mathematically advanced courses. Some students describe how the increased abstraction led to a decrease in motivation for their studies; for example: 'I've lost my motivation this year – it's just theory,' and 'There's something missing in terms of what is happening this year – I've lost that 'Oomph' in Maths.'

Secondly, *interactive engagement* was a key aspect in the first-year ECP teaching that was less common in much of the second-year teaching. The semantic profile in Figure 11.3 shows that student engagement was a key feature of the lecture sequence: the line thickness coding on the semantic profile indicates the many times during the lecture when there was student engagement. The faster pace of the second-year courses precluded much interaction with lecturers. Students noted how they missed this engagement and would have welcomed more structured group work in their second-year classes; for example: 'Our lecturers are not interacting with us....They are so fast, they are just running with the notes' and 'It would be much better if we could work in groups, like in first year, because you can work with someone else than working on your own. It was more effective.'

Thirdly, the range of *representational modes* used in the second year narrowed. While ECP Physics teaching explicitly incorporated a range of representational modes (gestures, diagrams, graphs, mathematical equations), in the second-year courses, mathematical representations inevitably became more prevalent. This is to be expected in senior Physics and Mathematics courses. However, what the second-year Mathematics students noted was that representations with strong semantic density were often taken for granted and not explicitly unpacked in the teaching; for example: 'The problem is, now everything is abstract. We have to picture these problems. I struggle to visualize them. I tried to, but you *have to capture all these concepts visually*,' and 'Our lecturers teach us how to draw graphs but never teach us *how to view them*…. I have a lot of sketches in my notebook that I still don't understand.'

In summary, the Semantics analysis highlighted the discontinuities in teaching practice from first year to second year. These included: a curtailed semantic range concentrated more at the abstract level, less interactive engagement, a narrower range of representations used and less explicit unpacking of these representations in second year.

Specialization analysis

Data from student interviews was analyzed with the Specialization concept of 'knowledge-knower structures,' which highlights the importance not just to focus on disciplinary knowledge but also 'knower dispositions.' Perceptions of Mathematics as an élite code were evident in interviews with the second-year Mathematics students, where the notion of 'innate talent' was implicit. One student commented that most of the Mathematics postgraduate students seemed to come from outside of South Africa and questioned whether local students were perhaps not strong enough for postgraduate Mathematics studies.[1]

Furthermore, student interviews suggested that some of the learning-context knower dispositions that had begun to be developed in the ECP were no longer explicitly addressed in the second year Mathematics. These included encouraging students to work independently, to work collaboratively on whiteboards and discuss Mathematics. Students noted that they missed the opportunity for structured group work; for example: 'We are not interactively doing the work in class, most of us we are doing the work at home alone. I feel like we should do group work.'

Ellery notes that 'while disciplinary knowledge tends to form the main focus of science courses, becoming and being an independent learner is usually expected of students but is seldom explicitly articulated nor specifically supported, and therefore remains part of the "hidden curriculum"' (Ellery 2019: 234). She argues that knower dispositions, such as becoming an autonomous learner, needs to be explicitly modelled and scaffolded.

Students' reflections on their experience of second year suggested that they found the abrupt lack of this modelling and scaffolding of their independent learning difficult. It was assumed that students would work through notes and exercises at home, but this was not made explicit nor guided; for example:

> When you advance to second year maths you just get a shock. This year, in second year maths, the lecturer just reads the notes and explains a few concepts and just – you need to do it all at home. There's no time [as in

first year] that you have to work on something for weeks, it's just about what you are doing at home.

I feel like there's lots of gaps this year. You have to constantly go back, which is, you have to do the stuff everyday in order to get it. But the more you go back, the more you fall behind and the more you create more gaps for yourself. Unless you can work very fast, your time will fall short. This year, it is all about how you use your time. In first year, it wasn't like this. When we started with second year, it was like 'Boom!' They were all throwing things on us, it is so overwhelming.

In summary, the Specialization analysis highlighted the prevalence of an élite code perception among students, as well as a lack of modelling and scaffolding of learning-context knower dispositions in the second year Mathematics.

Rethinking teaching practices based on LCT analysis

The LCT tools had provided useful insights into the challenges students face in the transition to second year. The Semantics findings suggested that attentiveness to particular aspects of the teaching (semantic range, interactive engagement and the use of multiple representational modes) would be likely to support students in accessing the disciplinary knowledge and in navigating the 'epistemic transition' to second year. The Specialization findings suggested that attentiveness to the knower dispositions needed for STEM studies would also be important.

The next step was to use this LCT analysis to rethink and redesign aspects of the second-year teaching and curriculum. The high failure rate in second-year Mathematics was a grave concern, and so the first author (the faculty's teaching and learning specialist) presented the research findings to the Mathematics Department. This generated much discussion among the second-year lecturers and a willingness to work alongside the teaching and learning specialist in rethinking the second-year courses. LCT provided a useful conceptual framework for this collaboration; as Clarence has noted in her work with academic staff, the LCT tools assisted 'both academic development practitioners and disciplinary educators, working collaboratively, to analyse and change pedagogical practice in higher education' (2016: 126). This model of collaboration between educational specialist/academic literacy practitioner and disciplinary lecturers is described by Jacobs (2007), who argues that disciplinary lecturers are so immersed in their respective disciplines that the representations and discourse features of their discipline tends to be tacit and often taken for granted and that they may therefore find it difficult to make these discipline representations explicit to their students (see also, Marshall *et al.* 2011, for an example of this collaborative model in the context of Physics). In the section that follows, we discuss how the findings from the LCT analysis were used to rethink teaching practices.

Changes introduced on the basis of Semantics analysis

The purpose of classroom sessions was altered – instead of class time being used for the transmission of course material, students were expected to come to class prepared so that time could be spent tackling Mathematics tasks in class. This was achieved by replacing the traditional lecture format with an interactive workshop format, encouraging *student engagement*, discussion and 'talking Mathematics.' As one student who had failed the course in the previous year commented: 'Last year we would just sit in rows and listen; now we can interact and talk Mathematics to each other.'

There was a deliberate focus on widening the *semantic range* in classroom sessions through referring to specific examples whenever feasible. In Mathematics, the process of moving from specific examples to the general principles is termed 'abstraction,' which many students find challenging. Wiggins (2018) argues that Mathematics lecturers need to emphasize that Mathematics has its roots in the study of real-world problems and to demonstrate abstract concepts through specific examples whenever feasible.

Mastering disciplinary representations takes place over an extended period of time, beyond first year. Yet, as Fredlund *et al.* (2012) note, lecturers are often so familiar with disciplinary representations that are oblivious to the 'learning hurdles' involved in interpreting the intended meaning of these representations (also see, Conana *et al.* 2016, 2020). In response to the students' and researchers' observations that representations were often taken for granted in the second-year Mathematics course, there was also a more explicit focus on exploring and unpacking a *range of representations.* Wood *et al.* (2007) argue that a key purpose of undergraduate Mathematics teaching is 'to assist students to make links between various representations of mathematical concepts' (p. 12), including oral and written language, mathematical notations and visual diagrams. They argue that these links between representations 'form the basis for deep learning and fluency in working with mathematical ideas' (ibid.:12).

Changes introduced on the basis of Specialization analysis

The Specialization analysis emphasized that, while Mathematics knowledge is central (relatively strong epistemic relations), more time is needed in the curriculum to address knower dispositions (strengthening social relations). As Ellery (2019) argues, in many traditionally content-dominated STEM courses (with a strong knowledge code), there is not enough explicit focus on developing the knower dispositions, values and ways of thinking important for success in the discipline.

Drawing on Ellery's work, the interventions were designed to develop students' knower dispositions. One key aspect of the interventions was to challenge the 'innateness' belief that situates Mathematics as an élite code in students' eyes. Through developing students' dispositions, such as confidence, autonomous learning, deep approaches to learning and enhanced metacognitive capabilities, the notion was emphasized that success in Mathematics relies on 'attitude, not aptitude.'

Before the start of the second year, a weeklong 'boot camp' was held for all second-year Mathematics students. This was explicitly geared towards supporting students in the transition to second year: sessions focused on overtly articulating the sorts of knower dispositions needed for second year Mathematics and these were modelled during the weeklong intervention.

Sessions on the nature of mathematical thinking developed the *production-context* knower dispositions relevant for Mathematics, the epistemic norms and ways of thinking in Mathematics. These included sessions on logic, and the role of proof in Mathematics. Hodds *et al.* (2014) recommend an intervention that focuses on logical relationships and introduced a pedagogical technique called *'self-explanation training'* to help students with the comprehension of proofs. Worksheets were developed that systematically deconstruct the theorem, followed by a rigorous set of step-by-step guidelines through the body of the proof that helps a student to understand technical terms used and to develop their cognitive capacity to handle the details of deductive arguments.

Many students initially voiced anxiety at the prospect of dealing with the advanced Mathematics of second year; the boot camp provided a broad, conceptual introduction to the second-year mathematical courses (Advanced Calculus and Linear Algebra), emphasizing the real-world origins of these mathematical fields and how students' high school and first-year Mathematics knowledge formed the foundation for these more advanced-level courses. In addition, the role of Mathematics as the language of the sciences was foregrounded (with sessions on Mathematics for Physics, Statistics, Computer Science and Chemistry by lecturers from these disciplines).

Other sessions addressed the *learning-context* knower dispositions appropriate for learning Mathematics. These included students' dispositions such as autonomous learning, adopting deep approaches to learning and developing metacognitive capabilities. Boot camp activities developed students' metacognition, with many opportunities to reflect on their learning and identify challenges. Mathematics education research shows the value of this sort of explicit focus on students' approaches to learning and their conceptions of Mathematics: Wood *et al.* (2012) identified three levels of conceptions of Mathematics, ranging from fragmented conceptions of Mathematics as a collection of components and techniques (level 1), Mathematics as a focus on models and abstract structures (level 2) and Mathematics as tools for understanding the world (level 3). Studies show that fragmented conceptions of Mathematics (level 1) are linked to surface approaches to learning and poor-quality learning outcomes, whereas the more cohesive conceptions of Mathematics (levels 2 and 3) are linked to deep approaches to learning and better learning outcomes (Crawford *et al.* 1994, 1998).

Besides the initial boot camp, the teaching approach of the second-year Mathematics courses was adapted in response to the research findings. As noted above, the traditional lecture format was transformed into a workshop format. Students' capacity to *work independently* was supported and developed through assigning class preparation worksheet tasks so that more time was freed up during class for in-class activities and discussion of the challenges the students encountered in the tasks.

Students' *metacognition* was developed through activities that encouraged them to reflect on their approaches to learning and identify problem areas they were experiencing. In class, students worked in groups on tasks and were encouraged to discuss their thinking, and compare solutions. Students were encouraged to present their struggles to the whole class, and peers would then suggest strategies and different approaches to solving the problems.

The development of these knower dispositions was enabled by the sense of a classroom community, which gave students the space and confidence to 'speak' Mathematics and to feel part of a learning community (see Engstrom and Tinto 2008, on the impact of learner communities on student retention in higher education). This minimized the Mathematics anxiety many students were laden with. In addition, physical space was created in one of the Faculty buildings for students to meet informally and discuss Mathematics – this space had movable tables and chairs, and plenty of whiteboards for working on tasks together. During some classes, postgraduate students were invited to do short twenty-minute presentations; these students were role models for second-year students, helping them to envisage themselves as becoming mathematicians.

Impact of the intervention on student learning

Overall, the Mathematics lecturers reported an increased confidence in students and improved attitude towards Mathematics learning. The opportunity to 'talk Mathematics' built their confidence and developed their sense of themselves as 'Mathematics students.' Their sense of agency in relation to their Mathematics learning developed; as their metacognitive approach developed, they became less focused on getting the right answer and more focused on the mathematical process. This was evident in the student exchanges on an informal 'WhatsApp group' the students had created: the emphasis was not on sharing solutions to problems, but rather on providing feedback to each other on approaches to problems. Examples of peer comments were: 'Did you think about …'? 'What am I missing here?'; 'No-one post a solution please – I want to figure it out myself.'

As students' sense of identity as Mathematics students deepened, there were unexpected developments: the students formed a Mathematics Club (which arranged lunchtime seminars and events), and a Mathematics Hub (a campus residence-based Mathematics Club). Students' confidence to speak about Mathematics also led to the establishment of an outreach project in local high schools. More students wanted to take part in the South African Tertiary Mathematics Olympiad, indicating that the science knower disposition of curiosity (see Ellery 2018) had been fostered in these students, as they enjoyed applying their creativity and critical thinking skills, and exploring concepts beyond the second-year curriculum.

The students also began to reflect more on the relationship between Mathematics and their other second-year science courses, and lecturers in these other courses noted improved success in these courses. As much as motivation, attitude and confidence matter, the main currency of undergraduate education is assessment results, and here the impact of the intervention was significant: the pass-rate in the second-year Mathematics courses increased significantly, from about 30% to about 80%.

Another interesting development was that the positive 'turnaround' in second-year Mathematics had a wider impact in the Mathematics Department as a whole. More senior postgraduate tutors were now willing to tutor second-year courses: in the past, they had found tutoring demoralizing, but now they were keen to work with these motivated students and keen to motivate them further. Similarly, the newly motivated second-year students became keen to tutor the first-year students, and as a result, the first-year Mathematics pass-rate also increased. In the year subsequent to the first implementation of this intervention, the enrolment in third-year Mathematics more than doubled, from 15 to about 40 students.

Concluding remarks

This research on the transition from first to second year in higher education addresses the paucity of research in this area. We note that in South Africa, as elsewhere, most of the research focuses on the transition from school to higher education and neglects 'epistemic transitions' (CHE 2013) later in the trajectory of undergraduate students.

Concepts from Semantics and Specialization provided useful insights into the challenges students face in the transition to second year. The Semantics analysis suggested that attentiveness to particular aspects of the teaching (greater semantic range, more interactive engagement, the use of multiple representations and more explicit unpacking of these) would be likely to support students in accessing the disciplinary knowledge and in navigating the 'epistemic transition' to second year. The Specialization analysis highlighted the way that Mathematics operates as an élite code for many university Mathematics students; the findings suggested that, while a focus on knowledge tends to dominate undergraduate Mathematics teaching, attentiveness to the *production-context* knower dispositions and *learning-context* knower dispositions (Ellery 2018) needed for success in Mathematics studies would also be important. These LCT research findings were then used to frame an educational intervention in second-year Mathematics. This intervention was found to lead to significant changes in students' attitudes towards Mathematics learning, as well as in their learning outcomes. Although the focus of our analysis in this chapter was second-year Mathematics, these findings would likely be applicable to a range of STEM disciplines. The Semantics and Specialization tools are valuable for teasing out ways in which the semantic features and the knowledge-knower structures of a discipline might be made more accessible to students.

Note

1 Under Apartheid in South Africa, deliberate education policy restricted access to quality Mathematics education for black learners.

References

Ashcraft, M. H. (2002) 'Math anxiety: Personal, educational, and cognitive consequences,' *Current Directions in Psychological Science*, 11(5): 181–185.

Bernstein, B. (1996) *Pedagogy, symbolic control and identity: Theory, research, critique*, Taylor and Francis.

Bernstein, B. (2000) *Pedagogy, symbolic control and identity: Theory, research, critique*, Revised edition. Rowman and Littlefield.

Bing, T.J. and Redish, E.F. (2009) 'Analyzing problem solving using math in physics: Epistemological framing via warrants,' *Physical Review Special Topics – Physics Education*, 5: 020108.

Blackie, M. (2014) 'Creating semantic waves: Using Legitimation Code Theory as a tool to aid the teaching of chemistry,' *Chemistry Education Research and Practice*, 15: 462–469.

Boaler, J. (2016) *Mathematical mindsets: Unleashing students' potential through creative math, Inspiring Messages and Innovative Teaching*, John Wiley and Sons.

Bourdieu, P. (1994) *In other words*, Polity Press.

Clarence, S. (2016) 'Exploring the nature of disciplinary teaching and learning using Legitimation Code Theory semantics,' *Teaching in Higher Education*, 2(2): 123–137.

Conana, H., Marshall, D. and Case, J. (2016) 'Exploring pedagogical possibilities for transformative approaches to academic literacies in undergraduate physics,' *Critical Studies in Teaching and Learning*, 4(2): 28–44.

Conana, H., Marshall, D. and Solomons, D. (2019) 'Legitimation Code Theory as a theoretical lens to think about transition,' *Alternation*, 26(2): 183–212.

Conana, H., Marshall, D. and Case, J. (2020) 'A semantics analysis of first year physics teaching: Developing students' use of representations in problem-solving,' in C. Winberg, S. McKenna and K. Wilmot (eds) *Building knowledge in higher education: New directions in teaching and learning*, Routledge.

Council on Higher Education (CHE) (2013) 'A proposal for undergraduate curriculum reform in South Africa: The case for a flexible curriculum structure.' Report of the Task Team on Undergraduate Curriculum Structure. Retrieved from: http://www.che.ac.za/media_and_publications/research/proposalundergraduate-curriculum-reform-south-africa-case-flexible.

Crawford, K., Gordon, S., Nicholas, J. and Prosser, M. (1994) 'Conceptions of mathematics and how it is learned: The perspectives of students entering university,' *Learning and Instruction*, 4(4): 331–345.

Crawford, K., Gordon, S., Nicholas, J. and Prosser, M. (1998) 'University mathematics students' conceptions of mathematics,' *Studies in Higher Education*, 23(1): 87.

De Freitas, E. and Sinclair, N. (2014) *Mathematics and the body: Material entanglements in the classroom*, Cambridge University Press.

Ellery, K. (2018) 'Legitimation of knowers for access in science,' *Journal of Education*, 71: 24–38.

Ellery, K. (2019) 'Congruence in knowledge and knower codes: The challenge of enabling learner autonomy in a science foundation course,' *Alternation*, 26(2): 213–239.

Engstrom, C. and Tinto, V. (2008) 'Access without support is not opportunity,' *Change: The Magazine of Higher Learning*, 40(1): 46–50.

Fredlund, T., Airey, J. and Linder, C. (2012) 'Exploring the role of physics representations: An illustrative example from students sharing knowledge about refraction,' *European Journal of Physics*, 33: 657–666.

Georgiou, H. (2014) 'Putting physics knowledge in the hot seat: The semantics of student understandings of thermodynamics,' in K. Maton, S. Hood and S. Shay (eds) *Knowledge-building*, Routledge.

Hodds, M., Alcock, L. and Inglis, M. (2014) 'Self-explanation training improves proof comprehension,' *Journal for Research in Mathematics Education*, 45(1): 62–101.

Hunter, M.S., Tobolowsky, B.F., Gardner, J.N., Evenbeck, S.E., Pattengale, J.A., Schaller M.A. and Schreiner, L. (eds) (2010) *Helping sophomores success: Understanding and improving the second year experience*, Jossey-Bass.

Jacobs, C. (2007) 'Towards a critical understanding of the teaching of discipline-specific academic literacies: Making the tacit explicit,' *Journal of Education*, 41(1): 59–82.

Lamont, A. and Maton, K. (2008) 'Choosing music: Exploratory studies into the low uptake of music GCSE,' *British Journal of Music Education*, 25(3): 267–282.

Lindstrøm, C. (2012) 'Cumulative knowledge-building in a hierarchically knowledge structured discipline: Teaching university physics to novices,' Paper presented at the *Seventh International Basil Bernstein Symposium*, Aix-en Provence, France.

Ma, X. (1999) 'A meta-analysis of the relationship between anxiety towards Mathematics and achievement in mathematics,' *Journal for Research in Mathematics Education*, 30(5): 520–541.

Marshall, D., Conana, H., Maclons, R., Herbert. M. and Volkwyn, T. (2011) 'Learning as accessing a disciplinary discourse: Integrating academic literacy into introductory physics through collaboration partnership,' *Across the Disciplines*, 8(3): 1–9.

Marton, F. and Säljö, R. (1976) 'On qualitative differences in learning. 1 – outcome and process,' *British Journal of Educational Psychology*, 46: 4–11.

Maton, K. (2009) 'Cumulative and segmented learning: Exploring the role of curriculum structures in knowledge-building,' *British Journal of Sociology of Education*, 30(1): 43–57.

Maton, K. (2013) 'Making semantic waves: A key to cumulative knowledge-building,' *Linguistics and Education*, 24(1): 8–22.

Maton, K. (2014) *Knowledge and Knowers: Towards a Realist Sociology of Education*, Routledge.

Maton, K. (2020) 'Semantic waves: Context, complexity and academic discourse,' in Martin, J. R., Maton, K. and Doran, Y. J. (eds) *Accessing Academic Discourse: Systemic functional linguistics and Legitimation Code Theory*, Routledge, 59–85.

Maton, K. and Chen, R. T-H. (2016) 'LCT in qualitative research: Creating a translation device for studying constructivist pedagogy,' in K. Maton, S. Hood and S.

Shay (eds) *Knowledge-building: Educational studies in Legitimation Code Theory*, Routledge, 27–48.

Mtombeni, T. (2018) 'Knowledge practices and student access and success in General Chemistry at a Large South African University,' Unpublished PhD thesis, Rhodes University.

Wiggins, H.Z. (2018) 'Mathematics: Forget simplicity, the abstract is beautiful - and important.' *The Conversation*. https://theconversation.com/mathematics-forget-simplicity-the-abstract-is-beautiful-and-important-91757. (Accessed 27 May 2021).

Wolff, K. (2015) 'Negotiating disciplinary boundaries in engineering problem-solving practice,' Unpublished PhD thesis, University of Cape Town.

Wood, L.N., Joyce, S., Petocz, P. and Rodd, M. (2007) 'Learning in lectures: Multiple representations,' *International Journal of Mathematical Education in Science and Technology*, 38(7): 907–915.

Wood, L., Petocz, P. and Reid, A. (2012) *Becoming a mathematician*, Springer.

Yorke, M. (2015) 'Why study the second year?' in C. Milsom, M. Stewart, M. Yorke and E. Zaitseva, (eds) *Stepping up to the second year at university: Academic, psychological and social dimensions*, Routledge.

Part V
Science Education Research

12 Navigating from science into education research

Margaret A.L. Blackie

Introduction

The threshold into publishable education research for someone who is used to publishing disciplinary research in a STEM environment is non-trivial (Adendorff 2011). The purpose of this chapter is to provide something of a map. It will not provide a 'method' but rather give an explanation of the distinctions between science and social science. Not all education research is social science, but it is this aspect of education research that can appear to be a non-navigable wilderness to those who enter from a STEM discipline. The first part of the chapter is dedicated to a description of a philosophy called 'critical realism,' which offers a useful foundation from which we can view both research in science and in social science and thus show some of the similarities and distinctions between the two. This is followed by a brief discussion about Legitimation Code Theory (LCT). Finally, pointers are given on what is necessary to consider when undertaking a study enacting LCT.

Critical realism

The real, the actual and the empirical

Critical realism holds a realist ontology, whilst recognizing that knowledge of that reality is socially constructed. The 'realist ontology' means that the physical world is real and the mechanisms that account for change in the physical world are independent of humans. However, the physical world and humanity are such that we have a capacity to observe, interrogate and devise explanations for those mechanisms of change. This practice is what we call 'science.' The explanations of those mechanisms are, however, subject to two different kinds of limitations. The first limitation is that we can only describe or attempt to explain what we can observe. We do not know, and actually cannot know, how much of reality we can observe. Critical realism, therefore, divides that which is ontologically real into three realms (Figure 12.1): the real (the whole), the actual (where mechanisms actually operate) and the empirical (that which is observable by human beings). Thus, the scientific method as it is usually taught in undergraduate programmes is constrained

DOI: 10.4324/9781003055549-17

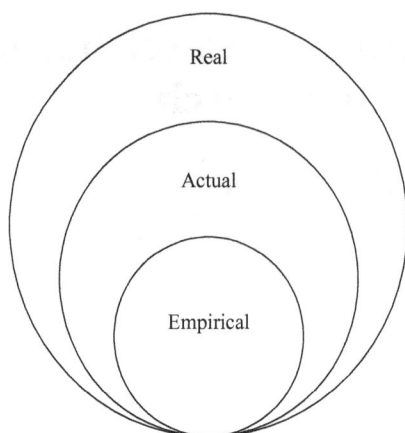

Figure 12.1 Nested relationship between the real, the actual and the empirical domains.

to the empirical (Bhaskar 1978). The second limitation is that the description or tested explanation given by science, the scientific concept, is often presented as a 'fact,' but there is a fundamental difference between the concept and the thing being described by the concept.

Two common errors made by some scientists are scientism and naïve realism. These are usually blind spots that are unconsciously fostered in our current dominant forms of science education and thus may be quite prevalent among academic scientists. 'Scientism' is the erroneous limitation of that which is knowable to that which is empirically accessible in the physical world. Rather than recognizing that science can only explain that which is observable. Thus it gives an inflated description of science and tends to overlook the value of a knowledge area which does not conform to the norms of scientific observation. 'Naïve realism' is the failure to make a distinction between that which is observable and the description of the thing observed. In other words, the scientific concept is conflated with the object of study.

It should be pointed out here that for many scientists, there may be some discrepancy between their espoused theoretical framework (the framework they say they hold) and their operational theoretical framework (the framework they actually use). Few scientists would argue against the part of critical realism thus far described. However, few scientists need to make the distinction between that which is described and that which is real in their research. For example, there is no need to make a distinction between the concept of molecules (the description) and the actual entities that are reacting in the flask (that which is real). As a result scientists often fail to make this distinction explicit in their teaching. Thus, whilst their espoused theoretical framework may align with critical realism, their operational theoretical framework may be one of naïve realism. The student is therefore exposed to implicit naïve realism, rather than the more sophisticated position that the academic claims they hold. This is not necessarily an issue, but it does hold the possibility that a student could graduate with a Bachelor of Science degree and not recognize that there is a distinction between the concept of the molecule and the entity in the flask. Within many scientific disciplines, the operational theoretical framework is simply assumed to be common within the discipline and therefore is not made explicit in the scientific literature. As a consequence, many scientists simply have no need to give it much thought (the unconscious blind spot alluded to above). This situation is exacerbated by the fact that many of the significant scientific theories taught at an undergraduate level have been established for decades and thus the contingency of scientific knowledge is not made visible.

The scientist as agent

The second aspect of critical realism that is important for the academic scientist who is dipping their toes into education research is more likely to require a conscious shift in perspective. We are used to thinking of the scientist as the neutral observer. For Bhaskar (1978), the scientist is an agent in the process. In *A Realist Theory of Science*, Bhaskar (1978) makes an argument for the open nature of the real domain. That is to say, the real world is always an open system. The practice of scientific research, in general, is the intentional closure of a system in order to investigate a particular mechanism – the scientific method. Thus, the scientist is not merely a passive observer but also an agent who actively closes a system in order to investigate a particular empirical phenomenon, of a particular actual event, caused by a particular mechanism or confluence of mechanisms. What makes science reproducible by different scientists in different parts of the world is the reality of the underlying mechanisms. Once the conditions of closure to investigate a specific mechanism or set of related mechanisms are adequately described (the experimental method), the experiment can reliably be expected to be reproducible by a second scientist. The understanding of the mechanisms is 'scientific knowledge' and is a product of society. It will thus be subject to influence by history and personal experience. Over time, this particularity of the first description will be refined as more data is brought to bear on the mechanism and as the science itself evolves. Nonetheless, the existing models and explanations of the mechanism (scientific knowledge) will profoundly influence the manner in which both the experiment is carried out and the way in which the data is interpreted.

It is probably helpful at this point in the discussion to pause and consider a particular example from your own field of research. I am a synthetic chemist, so a typical experiment in my field would be designing a reaction to form a specific product. Presuming the product has not been reported before, I will draw on various previously described reactions to design the new synthesis. I will also draw heavily on my experience of performing reactions. If I have a choice, I will choose a reaction I am familiar with over one I have never done before. Once the reaction has been designed, I will try it out and repeat several times until I am satisfied that it is reliably reproducible. Note that the design of the experiment is guided significantly by my own prior experience and my understanding of what methods can theoretically be brought to bear on this particular reaction.

In more general terms, then, the experiment once reported should be reproducible, but the design of the initial experiment is subject to variation based on the individual researcher and will be substantially influenced by the theoretical understanding of that researcher at that point in time. What makes science 'objective' is that real mechanisms are at work in a well-defined, closed system. The chemical reaction in the previous paragraph works because there is a real mechanism that is at work and which I have harnessed to create the molecule that I desire. If the system is sufficiently well described, it can be reproduced by another person in a different environment. Someone in Cairo could follow my description and should get the same product. The 'objectivity' is thus a product of the well-defined, closed system, not a personal quality of the scientist. Note though that the design of the first experiment to investigate the existence or nature of a particular mechanism will be subject to influence by the personal experience of the researcher. This is Bhaskar's point that scientists are not passive observers but rather active agents in the production of this activity called science (Bhaskar 1978). Too often, the reproducibility of the experiment is conflated with a presumed objectivity on the part of the scientist. These two things are separate and not reducible to one another.

It is this element of critical realism which is likely to cause most discomfort to the vast majority of academic scientists who deeply value the objectivity of science. Once again, scientific knowledge is objective, but scientists may not be. Whilst this shift in perspective may be a little uncomfortable to begin with, it allows scientists to explore the practice of science as a deeply creative process. This process must be connected to the real, through the empirical observation of actual events or through the precipitation of actual events brought about by the conditions of closure to create particular empirical outcomes. Thus, as scientists, we must continue to strive to describe what we have done in ways that others can reliably reproduce the work. However, exploration of new intellectual spaces is deeply personal and deeply creative.

As a brief aside, it is worth recalling that the use of the passive voice is prevalent in publications in the natural sciences. Cooray (1967) claims that the passive 'helps the writer to maintain an air of scientific impersonality.' Banks' (2017) study on the use of the passive voice in scientific writing from 1985 to 2015 indicates that the use of the passive voice is declining. Banks makes this note in the paper:

> Active voice with a first person plural subject tends to be used when the authors wish to underline a personal contribution, while passive is used for established or standard procedures. Where a contrast is made, authors tend to use the active voice for their own work and the passive for the work of others. And authors use the passive for speculating on their own future work.

> (Banks 2017: 12)

This distinction between the use of the passive voice when describing an established procedure and using the active voice when describing new actions taken by the researcher is directly in line with raising the visibility of the scientist as an active agent.

Emergence and closure

Having established, then, that the real domain is open and that scientific research is the process of intentionally closing a system in order to observe a particular empirically observable event, it follows that not all systems may be perfectly closable. The degree of closure possible in the system is one of the variables in research of all kinds. It is related to the degree of separation between the human person and the aspect of nature under investigation. There are a few notable exceptions to the possibility of closure of the system in natural science. The first is Astronomy. When objects in the cosmos or indeed the entire cosmos are the focal point, any attempt to close the system is not feasible. But the difference in scale between any human action and the mechanisms being empirically observed mean that this is not likely to be a problem. The second exception is more complex. At the quantum level, we have evidence that the presence of the observer alters the outcome. However, in the middle ground where most scientific research is situated, it is possible to close the systems under observation sufficiently. Where there is a question of potential attainable degree of closure, 'control experiments' are used.

Critical realism is an emergentist philosophy. This means that it allows for complexity to give rise to new mechanisms; for example, the behaviour of cells may be explained on the basis of some molecular interactions but is not reducible to those molecular interactions. That is to say, if one had a full understanding of molecular interactions, the behaviour of cells would not be entirely predictable from that data set. In other words, the behaviour of cells is an emergent property that is dependent upon molecular interactions but is not entirely reducible to molecular interactions. Thus, Biology is related to and built upon Chemistry but is a field in its own right and cannot be entirely reduced to Chemistry. Thus there are real mechanisms that exist at the level of cellular interaction which are not reducible to the level of molecular interaction. This is 'emergence'.

This concept of emergence then provides a bridge into the social world. We have just seen that cellular interactions cannot be reduced to molecular interactions, although they are dependent on molecular interactions. Likewise, the actions of an individual human being cannot be reduced to physical responses of the organism. More importantly, for education research, and indeed social science research, society has real mechanisms that cannot be reduced to the individual and that the social world is not entirely reducible to the physical world. Thus, in critical realism social structures are considered ontologically real. They are not unchanging in the same way as physical structures are, but they do have a reality that is irreducible to the individual. Examples of this include language, nation states or the banking system. These things exist as a product of human culture, and they shape the person born into that culture in particular ways. As an example, Boas, a nineteenth-century anthropologist, pointed out the phenomenon of 'sound blindness,' where researchers who had grown up speaking a European language were simply unable to hear differences in sounds made in some Pacific Island languages (Boas 1889). These languages use differences in tone to alter the meaning of words; to people who speak European languages, these differences are not noticed.

Here the distinction between social science and natural science comes into view. There are two significant differences. Firstly, in social science, closure of the system is substantially more difficult to achieve. The person of the social scientist is interrogating the behaviour of other humans who are consciously aware of human interaction and thus the social mechanism under investigation may be influenced by the fact of the study. In education research in particular where the researcher may also be the teacher, it is clear that the system is not closed. Secondly, the mechanism(s) under investigation may arise from cultural context. The implication of this is that one is unlikely to achieve the degree of reproducibility in education research that one can in science research. All one can do is to describe the social world sufficiently so that the mechanisms in action that are particular to the context may be less obscure.

Natural science and social science

The 'real'

In knowledge creation, there are three domains in relationship with one another – the real domain under investigation, the conceptual domain within which limits the stratum of the real which is observed and the community of researchers who contribute intellectually to the defined conceptual domain. These three domains exist in both social science and natural science. But, the degree to which the real domain can be closed varies, and the degree to which the human person is visible as an agent in knowledge creation varies. Nonetheless, the underpinning position is one of realism – that there is a 'real' domain that exists independently of the individual human person and that can be investigated.

One goal in natural science is to develop concepts that describe a phenomenon in the physical world. A second goal is to use those concepts to develop new technologies. In an analogous fashion one goal of social science is to develop concepts that describe a phenomenon in the social world. A second goal is to examine the ways in which social power operates in society. The exploration of the nature and dynamic of social power is thus a major focus of sociological research.

For a person coming from a background of natural science, there are two important conceptual elements which may not be immediately obvious. Firstly, that the social world does indeed have real mechanisms which give rise to events which are empirically observable. However, it is substantially more difficult to isolate and attribute cause unambiguously to a mechanism in the social world. Secondly, there are always social power dynamics in play. Thus, research into the description of those power dynamics is a legitimate form of knowledge-building activity. It is entirely possible to attempt to hold both goals in view at the same time, but it is more common for one goal to be favoured.

Knower-blindness

The impact of operational naïve realism on the practice of science is what Maton calls 'knower-blindness' (Maton 2014: 14). This is in contrast to the 'knowledge-blindness' which was prevalent in sociology of education literature in the 2000s (Maton 2014). The potential reality of knowledge was obscured. Thus, what was taught was de-emphasized in favour of developing the 'voice' of the student (Moore and Muller 1999). This can be understood as conflation of ontology (what is real) and epistemology (what is known) thereby reducing the real to what is known. All becomes epistemology, and there is nothing beyond the constructed concept. This position can be called ontological constructivism. This shift was prevalent in the social sciences and dominated the sociology of education in the 1980s and 1990s. This certainly influenced some science education too (Scerri 2003), but the impact was not felt significantly in tertiary science education. This may be because tertiary science educators have been largely ignorant of the science education literature until the push of scholarship of teaching and learning discourse became mainstream in higher education in the last decade.

In contrast, because scientists involved in tertiary education are also involved in scientific research, the position they tend to hold is naïve realism rather than ontological constructivism. As described above, natural scientists can tend to be blind to the influence of the social world on scientific research. Again there is a conflation between ontology and epistemology, but here the error is in the opposite direction. Epistemology is promoted to ontology: the concept is taken to be that which it was constructed to describe. Hence where social science erred towards knowledge-blindness, natural science erred towards knower-blindness.

The consequence of this is a lack of recognition of the significance of society on the propagation of science. A caricature of this was present in the response of some scientists to call for decolonization in the #FeesMustFall protests in South Africa. The position was clearly that science is inherently socially neutral because it is objective; therefore, there is no possibility of a decolonized science curriculum (Adendorff and Blackie 2020). This position is one of naïve realism. From this position, when one observes a social dynamic at play in education, the desire is to remove the social dynamic to retain the holy grail of objectivity in science education. This is reinforced by the use of the passive voice in the scientific literature mentioned above. However, critical realism would suggest that this move is a fool's errand.

As scientists begin engaging in education research, a major mental shift needs to happen. We must bring into view the reality of the social structure. Here, as previously mentioned, it is important to recognize that social structure also has an ontological reality that is irreducible to the action of the individual. One of the most influential authors on this point is Margaret Archer (2000), who, building from a critical realist starting point, argues for the importance of recognizing the impact of both structure (the level of society) and agency (the level of the individual) in effecting any kind of impetus to transformation.

Role of the concept

I have already indicated the problem of naïve realism where the distance between the conceptual world and the physical world is collapsed. In science, the conceptual world is a constructed world that has correspondence to the physical world. In the idealized notion of how science progresses, the concept shifts from a proposal to something that is generally accepted by the field. This process follows several steps. Initially, a scientist publishes a paper describing the observation of a particular empirical phenomenon. This phenomenon is then investigated by others and sooner or later a mechanism to explain the phenomenon is proposed. The limits of the mechanism are then explored, and the mechanism is refined. These refinements are in turn published. After another period of time, a single refined mechanism becomes favoured and is taken to be accepted knowledge. From this point on, two independent processes occur. Firstly, the refined mechanism subtly shifts from being an explanation for an empirical observation to being 'how the world is.' That is the distinction between the conceived mechanism and the real physical mechanism collapses. This is the slippage into naïve realism. Secondly, the refined mechanism becomes the conceptual foundation which shapes the way the scientist thinks. In this second sense, the concept does become real. Its existence has an influence on scientists working in that field. Concepts frame the way in which we approach our scientific enterprise. Having indicated that the scientists may not be the source of objectivity in science, it is important to acknowledge here that the acceptance of a new concept by the community of scientists is surely somewhat influenced by politics and personal power.

Teaching science

Most of the fundamental sciences are well established in that there is a broad, robust conceptual foundation. For many of the established sciences, there has been little change to this foundation in the last several decades. There are some exceptions – for example, developments in Molecular Biology are ongoing and continue to shape aspects of Biology. Nonetheless, there is usually general consensus on the conceptual foundation which forms the basis of many undergraduate science programmes. So science programmes across the world tend to have a common core.

At this point, it is helpful to introduce the idea of the 'epistemic–pedagogic device' (Maton 2014). This idea is built upon the foundation of Bernstein's 'pedagogic device' (2000). There are three interrelated fields of practice in education:

- The *field of production*, where 'new' knowledge is created (in science, this is often the research laboratory).
- The *field of recontextualization*: where knowledge from the field of production is selected, arranged and evaluated as curriculum and textbooks for use in teaching and learning. (In science, this is usually done by the authors of textbooks, although in some countries, professional bodies may play an active role in defining curricula.)
- The *field of reproduction*, where students are taught and learn a subject area (in science, this is the lecture theatre and the teaching laboratory).

If we are working from a position of naïve realism rather than one of critical realism, we may not notice that there is an active process of reproduction. In the case of most university courses in the sciences, where a textbook forms the foundation of a course, few of us will have any engagement in the field of recontextualization other than to make a choice of textbook. In hierarchical knowledge structures where concepts are strongly interrelated and build from a common foundation such as are present in many of the fundamental sciences, there may be relatively little choice around what is included and what is excluded from the curriculum. In addition, because we are interested in developing conceptual thinking rather than fostering a way of viewing the world, the choice of what is included and what is excluded is far less obviously subject to social power and political capital. However, the conversations around decolonization in the different faculties of South African universities show that this is indeed a little more complex than we might first imagine.

In science education though, it is useful to at least pause and notice that the field of production and the field of reproduction are separated from one another. The way in which we teach science can be remarkably different from the way in which we practice science. In some cases, the science that we teach can become so neatly packaged and internally referenced that it requires little experiential involvement from the student. In fact, in many cases, we inadvertently operate from a presumption that the student is a blank slate and we, as educators, are there to draw a good solid conceptual outline that the student can fill in. Alex Johnstone (2010), a powerhouse in chemistry education research, gave a scathing critique of Chemistry education in precisely these terms:

> We need to rethink a lot of what we teach. This does not imply that we have been teaching bad chemistry, but rather that we have been teaching inappropriate chemistry at the wrong time and in the wrong way. We have been presenting chemistry in a way contrary to what we now know and understand about learning.
>
> (Johnstone 2010: 23)

Thus, it may be that the way in which we teach science does not necessarily bear any relation to the way in which the student relates to the world. Our curriculum may be a beautiful conceptually coherent synthesis but if it fails to provide a bridge to the lifeworld of the student, the subject will remain disconnected and potentially inaccessible.

The important point here is that there is a significant shift in focal point between research and teaching. When we teach science, the conceptual domain is central but frequently fails to recognize that the construction of the conceptual domain is a fundamentally social activity. The process of passing knowledge on is infused with and embedded in society and culture even when we are teaching things that appear to be socially neutral like the structure of the atom. To fail to attend to the power and reality of the social is naïve at best and wilfully ignorant at worst.

There are many ways to improve our educative offering. In this book, we have focused on LCT, which is just one of the frameworks that can be used to achieve this end. Several dimensions of LCT are described in more detail in Chapter 1 of this volume, and various enactments of each dimension are illustrated in detail in Chapters 2–11.

Legitimation Code Theory

What is Legitimation Code Theory?

LCT is a realist theoretical framework which has its roots in the sociology of education. Karl Maton's (2014) *Knowledge and Knowers* is the foundational text of LCT and is the source of much of what is written in this section. LCT is built on several sources. One of these is the work of Basil Bernstein who had an interest in making explicit the ways in which language was used to create social boundaries. Bernstein's work coincided with the massification of higher education and was therefore concerned with revealing the 'codes' in order to give epistemic access to people who did not 'belong' (Bernstein 2000, 2003). LCT, developed by Maton, aims to make visible the 'rules of the game' of any social field of practice (Maton 2014). Education is one such field of practice, and Maton's explicit driver is that of social justice. If the rules of the game can be made explicit to all, anyone can learn how to play and be successful. In addition, in making the rules explicit, they can be critiqued and where necessary changed to create a better system.

We have found that LCT appeals strongly to many STEM-based academics who have an interest in STEM education because of the clear focus on knowledge. The various dimensions of LCT afford different ways in which teaching and learning can be explored. Each dimension is well bounded and well defined. Thus LCT can be used to illuminate particular facets of teaching and learning through careful choice of the dimension and development of an appropriate translation device. Producing robust publishable education research does require more than this, but engaging with LCT to improve teaching is a very powerful first step that can be carried out relatively easily, even for a newcomer to STEM education research.

LCT can be used to excellent effect in STEM environments to reveal the challenges of conceptual complexity required for mastery in the subject. Here Semantics and the epistemic plane have proved to be useful tools thus far. Semantics allows the exploration of the threshold to conceptual grasp by separating out complexity from abstraction (Maton 2014). Complexity is the degree to which knowledge is condensed into particular practices, terms or symbols. Abstraction is the power of the concept to explain multiple empirical observations, for example, the concept of an atom is used to account for a multiplicity of phenomena studied under the umbrella of Chemistry. The epistemic plane (Maton 2014) allows the separation between the methods deemed as legitimate and the objects of study. When one is trying to determine the structure of a molecule in Chemistry, the object of study is clearly defined, but there are many methods which may be applied to give the necessary information. Alternatively, if one is trying to master a particular analytical technique, the method is clearly defined but the object of the study could be any molecule. These two kinds of study both qualify as 'Chemistry' but vary in the degree to which method and object of study are constrained.

LCT can also be used to plan lessons and structure curricula. The concepts of semantic waves and autonomy tours are useful here. Thinking about moving strategically between simple/concrete meanings and complex/abstract meanings and back again (semantic waves) is important in facilitating cumulative learning (Maton 2014). Considering what elements of experience or other knowledges can be drawn in to facilitate learning of the subject you are focusing on (autonomy tours) is an important part of integrating knowledge (Maton and Howard 2018).

At another level, considering the purpose of the degree and the kind of formation one wants to achieve through a particular programme may be augmented by the use of Specialization (Maton 2014). Is the exclusive focus on epistemic acquisition, or is there an element of professional development also in play? Considering what is required and therefore what is desirable can have a significant impact on designing a more integrated, or at least a more intentional, curriculum.

LCT can be used for myriad analyses, as well as the shaping of teaching practice. Nonetheless, it is useful to bear in mind the purpose of its creation – to make visible the 'rules of the game' for what makes a knowledge claim legitimate – who can make the claim and how the claim needs to be structured (Maton 2014). As such, it is clearly not just about knowledge but also about knowers. Ultimately, the purpose of most educative endeavours is to induct a novice into the field such that they have the capacity to become an expert. It is important to note that LCT is designed to be used in a fractal manner – that is, it can be used at any level, but we must be realistic about the limits of the spectrum accessible by the students. For example, if we are analyzing semantic density in an introductory course, we need to think about the capacity achievable by the top student end of the course, not the level that we have as academics.

Translation device

The feature of LCT which defines how the limits of the spectrum and the understanding of the spectrum within the specific context of your study is called a 'translation device' (Maton and Chen 2016). A translation device features at least three components. Firstly, the axis label representing the organizing principle or concept from the specific dimension (e.g. semantic density on the semantic plane), and the possible variations in strength, will depend on the level of refinement required by the specific study. Often, four levels of strength are described ++, +, – and – – although this should not be taken to be normative and a number of variations is permissible. Secondly, each of these levels of strength is given a specific description associated with, or determined by, the data set in hand. Thirdly, an example from the data should be included. This will make the interpretation of the study by the reader substantially more accessible and makes the study reproducible. Each axis (organizing principle) requires its own translation device.

The development of a translation device is usually an iterative process (Maton and Chen 2016). When one looks at the data, the possible variations begin to emerge. Suggested definitions or descriptions of the various levels of strength are then proposed and the data analyzed and coded accordingly. Inevitably, some data will not quite fit, and so the definitions will need to be modified or redefined. The analysis and coding then needs to be done again. There may be numerous iterations before the final translation device is settled upon.

The process of developing the translation device is an important learning curve. It is probably helpful if the researcher expects to be surprised in this process. In other words, the researcher should be open to learning from the data. It is here that we encounter the unexpected benefit of conducting research in a partially closed system. As we are likely to be researching elements of teaching our own discipline, we may discover new ways of thinking about what we are doing which may influence how we teach in the future. The stance here is not one of the disinterested expert but the researcher/ practitioner/teacher who is willing to be shaped by the process of researching.

Conclusion

The purpose of this chapter has been to make visible the ways in which social science is related to natural science using critical realism as a framework within which to illustrate the distinctions. Many natural scientists will approach education research unconscious of the philosophical framework they are operating out of. The consequence is a desire for 'rigour' through approaches such as use of a control group or pre-test/post-test type methods. The recognition of the inherent partial closure of an education environment as opposed to the fully closed system possible in the natural sciences afforded by critical realism should make visible the fact that these approaches will not actually provide rigour. A well-considered description of the environment in which the study has been done and the intention of the researcher will be more useful than any attempt to artificially remove the particularity of the context.

It is likely that any scientist embarking on the journey of engaging seriously with education research literature will find this chapter quite dense itself. It is probably worth keeping it 'on file' and returning to it periodically over the first few years of the exploration. Learning how to navigate a new intellectual space is itself an iterative process, and the conceptual map provided herein will make much more sense against the scaffold that will begin to be constructed in one's own mind.

Perhaps the most important point raised herein is the recognition that as scientists we operate out of various presumptions. Calling these into question or indeed simply making them visible can illuminate our understanding of ourselves as scientists and can potentially impact how we teach science and make the task of engaging with education research a little easier. This process of illumination is at the heart of LCT. LCT seeks to make visible the implicit 'rules of the game' which are required to gain access to, and to produce, knowledge which is seen to have value (legitimated) within a particular field.

References

Adendorff, H. (2011) 'Strangers in a strange land–on becoming scholars of teaching,' *London Review of Education*, 9(3): 305–315.

Adendorff, H. and Blackie, M. (2020) 'Decolonising the Science Curriculum: When good intentions are not enough,' in C. Winburg, K. Wilmot, & S. McKenna (eds), *Building knowledge in higher education: Enhancing teaching and learning with LCT*, Routledge.

Archer, M. S. (2000) *Being human: The problem of agency*, Cambridge University Press.

Banks, D. (2017) 'The extent to which the passive voice is used in the scientific journal article, 1985–2015,' *Functional Linguistics*, 4(1): 1–17.

Bernstein, B. (2000) *Pedagogy, symbolic control, and identity: Theory, research, critique*, Rowman and Littlefield.

Bernstein, B. (2003) *Class, codes and control: Applied studies towards a sociology of language*, Psychology Press.

Bhaskar, R. (1978) *A realist theory of science*, Harvester Press.

Boas, F. (1889) 'On alternating sounds,' *American Anthropologist*, 2(1): 47–54.

Cooray, M. (1967) 'The English passive voice.' *ELT Journal*, 21(3): 203–210.

Johnstone, A. H. (2010) 'You can't get there from here,' *Journal of Chemical Education*, 87(1): 22–29.

Maton, K. (2014) *Knowledge and knowers: Towards a realist sociology of education*, Routledge.

Maton, K., & Chen, R.T. (2016) 'LCT in qualitative research: Creating a translation device for studying constructivist pedagogy,' in K. Maton, S. Hood, and S. Shay (eds), *Knowledge-building: Educational studies in Legitimation Code Theory*, Routledge.

Maton, K. and Howard, S. (2018) 'Taking autonomy tours: A key to integrative knowledge-building,' *LCT Centre Occasional Paper*, 1: 1–35.

Moore, R. and Muller, J. (1999) 'The discourse of 'voice' and the problem of knowledge and identity in the sociology of education,' *British Journal of Sociology of Education*, 20(2): 189–206.

Scerri, E. R. (2003) 'Philosophical confusion in chemical education research,' *Journal of Chemical Education*, 80(5): 468.

Index

'A Can of Bull' 174, 179
abstraction 193–194
academic discourse 148
academic literacy courses 150
academic literacy/literacies 152, 206
academic scientist 1–3, 228–230
active teaching and learning practices 169, 171, 173
adaptive reasoning 195
agency 22, 34, 36, 219, 234
Anatomy 131
application 172, 187
Archer, M. 234
assessment 2–3, 12, 16, 26, 33, 42–55, 59–60, 79, 136, 138, 145, 149, 151–152
Astronomy 230
autonomous learner 22
Autonomy 4, 12–14, 16, 23, 173, 175, 177, 188
autonomy codes 14, 22–24, 33, 173, 177, 181
autonomy pathways 26–27, 29, 31, 177–178, 181, 187–188
autonomy plane 13, 23, 24, 26, 176–177, 187
autonomy tour 177–178, 182, 187, 237
axiological condensation 84, 98
axiological-semantic density 98
axiomatic system 201

Barnett, R. 22
Bernstein, B. 3–4, 138, 194, 207, 209, 235–236
Bhaskar, R. 134, 228–229
Biggs, J. 33
binary relation 200
Biochemistry 3

Biology 2, 11, 15–17, 24, 131, 150–153, 157–158, 160–166, 176, 208–209, 231, 234–235
Blackie, M. 11, 17, 84
Boas, F. 232
born gazes 9
Boughey, C. 150, 162
Bourdieu, P. 3–4, 207
Burning Question 182–183

Cartesian mind-body divide 210
Case, J., Marshall, D. and Grayson, D. 149
case-based learning (CBL) 172
categorized modifications 156
Chavez, C. 26
Chemistry 2–3, 11, 13, 15, 25, 44–45, 63–91, 97–99, 133, 171–172, 176, 208–209, 211, 218, 231, 235, 237
Chimbganda, A.B. 150
Christie, H. et al. 35
Clarence, S. 216
classroom activities 152
Clouder, L. 116
code clash 6, 33–34, 133, 137
code shift 121–122
code theory 207
collaborative pedagogy 151–152
common words 154–156
compact words 66, 88, 154–155, 158, 164
complexity 2, 9, 12, 15, 42–44, 46, 63–79, 84–85, 87, 104, 129, 133, 140, 151, 156–159, 162, 231
complexity in images 65–67
complexity of meaning/s 63, 65–66, 133, 151, 153, 157, 159, 207

Conana, H., Marshall, D. and Case, J. 2, 44, 46, 212–213
concept (the role of) 234
conceptual domain 232, 236
conceptual framework 3, 216
conceptual tools 207
conceptual understanding 195
condensation 12, 44, 49, 75, 84, 151, 158, 163, 166
conglomerate elements 156
conglomerate properties 156
conglomerate words 88, 154–155, 158
Conley, D.T. 21
consolidated words 88, 154–156
constellations 12, 66–69, 71–72, 151, 165
control experiments 230
Cousin, G. 104, 109, 114, 118
critical realism 17, 227–231, 233, 235, 238–239
critical thinking 169, 174
'CSI-type' activity 172
cultivated gazes 9
cumulative knowledge-building 3, 11, 84, 139, 142, 152
cumulative learning 41–44, 59–60, 115, 131, 142, 172, 237
curriculum 134

De Freitas, E. and Sinclair, N. 210
De Morgan, A. 200
decontextualized conceptual learning 172
Dedekind, R. 200
degree of closure (of a system) 230
Devlin, K. 194
differentiated support 204
Dimopoulos, K., Koulaidis, V. and Sklaveniti, S. 65
disciplinary boundaries 139
disciplinary integration 140
discipline-specific 111, 161–162
discourse
'big D' Discourse 148–151
d/Discourse 148–151
'little d' discourse 148–151
features 216
discursive relations 6–7, 197–198, 200–203
Docktor, J.L. and Mestre, J.P. 41, 46
doctrinal insight 6, 197, 199–204
'down escalator' profile 209
Duval, R. 194

educational research 134, 231–232
élite codes 5, 107, 143, 197, 209, 132, 143
Ellery, K. 14, 21–22, 211, 215, 217
embedded 156
embedded modifications 156
emergentist philosophy 231
empirically observable 233
Engle, R.A. and Conant, F.R. 149
epistemic-pedagogic device 235
epistemic plane 6–7, 16, 197, 199, 237
epistemic relations 5–7, 9, 107–109, 112, 114–121, 132–133, 144, 196–197, 209, 210
epistemic relativism 134
epistemic-semantic density 12, 15–16, 66–67, 69–79, 85–86, 88, 91, 94, 96, 99, 151–156
epistemic transitions 206
epistemology 233
epistemological access 2, 4, 22, 164
epistemological condensation 12, 84–85, 151, 158, 163
everyday language 149
everyday words 153–156
exotic codes 13–14, 24, 30, 32, 34, 176, 178, 180–182, 184, 187
extended curriculum programme (ECP) 206–207, 211–215
Extended Degree Programme (EDP) 25, 153

field of production 138, 235
field of recontextualization 138, 235
field of reproduction 138, 235
first-generation students 206
first-year curriculum 152
formalism 197, 199–200, 203
Fredlund, T., Airey, J. and Linder, C. 217
Fredholm, A. 106

Garraway, J. and Reddy, L. 22
gate keeping 210
gazes 8
Gee, J.P. 148
generalist words 154, 156
Generation Z 169
Georgiou, H. 11, 44, 60, 121, 212
Gibbs, G. 33
Gilbert, J.K. 65
Goodman, B.E., Barker, M.K. and Cooke, J.E. 171, 173–174
graduate attributes 169

hierarchical knowledge structures 209
higher-order thinking 149
high-stakes reading 150, 165
high-stakes writing 150, 165
Hilbert programme 201
Hockings, C. et al. 34
horizontal knowledge structure 194, 209
Human Biology 131–143
Human Life Sciences 174
Hung, W. 130
'HWW' principle 182

identity 8–9, 106, 116, 118, 202, 219
information-overload 171
innate ability 210–211
innate talent 215
innateness 217
insights (epistemic plane) 6, 197,
    199–204
insight shifting 198
interdisciplinary curriculum 131
interdisciplinary field 187
integration (of knowledge) 131,
    171–172
integrative knowledge-building 3, 16
interactional relations 8, 9
introjected codes 13, 24, 29–30, 32,
    34, 176, 178, 182

Jacobs, C. 150, 216
Johnstone, A.H. 65, 76, 78, 82–83,
    99, 235

Kelly-Laubscher, R.F. and Luckett, K.
    2, 149, 151
Kirk, S. 11, 150
Klement, B., Paulsen, D. and Wineski,
    L. 130–131
knower-blindness 233
knower codes 5, 107, 132, 196
knower/no insight 7, 197, 199–201,
    204
knowing that 196
knowing why 196
knowledge-blindness 4, 83, 134, 233
knowledge codes 5, 107, 132, 196, 209
knowledge-knower structures 196
knowledge structures 142–143, 207,
    209, 235

Land, R. 106, 114–116, 119
learning-context knowers 211, 218, 220
learning hurdles 217
Lee, O. and Fradd, S.H. 149

legitimate knower 8, 209
Legitimation Code Theory (LCT) 2–3,
    42, 65, 84, 103, 106, 131, 151,
    173, 196, 206, 227, 236
Lemoine, M. and Pradeau, T. 170
Lindstrøm, C. 212
located modifications 156
lying in 156

Marshall, D. and Case, J.M. 148–149
Mason, J. 194
math brain 210
mathematical activities 199, 204
mathematical activity 200–201
mathematical objects 193–194,
    199–202, 204
mathematical operations 194–195
mathematical proficiency 149, 193,
    195, 197–204
mathematical reasoning 193–194, 203
mathematical representations 193, 199,
    201–202, 204
mathematical structures 199, 202
Mathematics 13, 16–17, 24, 44, 53–54,
    104, 193–220
Mathematics ability 210
Mathematics anxiety 210
Maton, K. 3, 22, 43–44, 99, 107–108,
    150, 173, 211
Maton, K. and Chen, R.T. 178
Maton, K. and Doran, Y.J. 46, 85, 87,
    153, 156
Maton, K. and Howard, S. 4, 22, 24,
    26, 34, 177
Matruglio, E. 11, 84, 89
meaningful learning 173
metacognitive activity 149
metacognition 219
Meyer, J.H.F. and Land, R. 103–104,
    109, 111, 114
modifications 156–157, 165
Morrow, W. E. 4
Mortimer, E.F. and Scott, P.H. 98
Mouton, M. 2, 11
Mouton, M. and Archer, E. 2, 11
Mtombeni, T. 211
Music 210

National Centre for Case Study Teaching
    in Science (NCCTS) 174
Naïve realism 228, 233–235
newcomers 148, 158
nominalizations 208
nuanced words 88, 90, 154–156

object of study 132
objectivity 148, 211, 230, 233–234
occupational therapists 135
Occupational Therapy 135
one-way trips 177
ontic relations 6–7, 197–203
ontological relativism 134
ontology 233
organizing principles 207
Osborne, J. and Dillon, J. 1
outsiders 148

Paget, T. 35
pedagogic device 207, 235
pedagogy 1–2, 11, 14, 35, 150, 197
pedagogy of engagement 169
Peirce's logic of relatives 200
Perkins, D. 113
physical world 232
Physics 2, 11, 13, 15, 25, 28, 41–46,
        55, 59–60, 104–105, 110, 112,
        115, 118, 121, 171, 206–209,
        211–214, 218
Physiology 131, 169–171
Physiotherapy 135
physiotherapists 135
plain words 154–156, 162
positional autonomy 12–13, 23–24, 27,
        29–32, 175–179, 182, 185
power composition 159, 162–163
power language 150–152
power words 148
'powerful knowledge' 130, 145
principled knowledge 133
problem-based learning 130, 142, 169
problem solving 149, 172–173,
        177, 181
procedural fluency 195
procedural scientist (kind of knower) 140
production-context knowers 211,
        218, 220
productive disposition 195
professional knowledge 131
proficiency gaps 204
projected codes 13, 23–24, 28, 176,
        180–187
propositional learner 140
prosaic codes 10, 43, 53–54
proxy words 88, 156
purist insight 7, 197, 199–202, 204

Radiation Physics 103, 105, 107–110,
        117–122
range of representations 217
rarefied codes 10, 43, 54

real domain 231–233
realist ontology 227
real-life scenario 182
real mechanisms 232
real-world problems 171, 173
recontextualization 152
Reddy, S. 142
Reddy, S. and McKenna, S. 130
reflective learning portfolio (RLP)
        22–26, 30–36
relational autonomy 12–13, 23–24, 27,
        29–32, 175–179, 182
relativist codes 5, 107, 109, 132, 197
repacking 164, 166, 208, 212
representations of physicality 194
representation(s) 193–195, 214
representational modes 214, 216
reproducibility 230
return trips 177
rhizomatic codes 10, 43, 53–54
Riemann sum 201
Rootman-Le Grange, I. and Blackie,
        M.A.L. 84
Rowbottom, D.P. 106
'rules of the game' 4, 22, 150, 152,
        178, 236–237, 239
'Running Question' 174, 184–187

scaffold 157
Schreiner, L. and Hulme, E. 35
science classroom 13–14, 83, 97, 149
science education 1–2, 10–12, 14, 17,
        172, 228, 233, 235
science education research 1–2, 14, 17,
        83–84
science educator 1, 3, 17, 233
science pedagogy 2, 16
science student/s 3, 15, 36, 148–150
science talk 165
science teaching 9–10, 97
scientific communication skills 152
scientific d/Discourse 149–153
scientific language 148
scientific method 148, 229
scientific research 229, 233
scientific vocabulary 166
scientific writing 149–152, 165, 230
scientific writing skills 152
Scientism 228
second-year 206–207, 211–220
semantic codes 10–11
semantic density 9–12, 15, 43–60,
        65–66, 84–85, 87–89, 97, 133,
        151–153, 157, 164, 207–208,
        211–212, 214, 238

semantic gravity 9–11, 43–46, 48–50, 52–60, 84, 133, 140, 207–208, 211–212

Semantics 2–4, 9, 11, 16, 42–46, 59–60, 65, 84, 132–133, 151, 175, 207, 211–217, 220, 237

semantic plane 10–11, 43, 53–58, 208, 238

semantic profile(s) 85, 87, 89–90, 92, 95, 99, 134, 157, 164, 204, 208–209, 211–214

semantic profiling 208

semantic range 217

semantic shifts 152

semantic wave 11, 84, 92–93, 97, 99, 164, 237

semiotic representation systems 193

Shay, S. 44

situational insight 6, 197, 199–202

social gazes 8

social justice 131

social plane 8

social power 232, 235

social relations 2, 5–8, 83, 107–109, 112, 117, 119–122, 132–133, 144, 196, 209–211

social science 231–233

social world 134, 233

sovereign codes 13, 23–24, 27–32, 34, 176, 178, 181–184, 186–187

specialist words 156

Specialization 3–5, 15–16, 103, 106–108, 111, 114, 116, 120–122, 132–133, 175, 196–197, 203, 207, 209, 211, 215–217, 220, 237

specialization codes 5, 9, 107–108, 121, 132–133, 144, 196, 211

specialization plane 5–6, 107–108, 119–120, 132, 196, 209–211

specialized language(s) 110, 148–149, 194, 209

specialized vocabulary 162

STEM 17, 203, 206–207, 209, 211, 216, 217, 220, 227

student engagement 33, 35–36, 130, 214, 217

subject of study 132

subjective relations 8, 9

summative assessment 151–152

'surface' learning 170

stays 177

strategic fluency 195

strong grammars 194

structure of the knowledge 141, 144

Tan, K.C.D. et al. 118

target of study 175

technical-compacts 69, 71, 73, 77–78

technical-conglomerate(s) 69, 71, 74–75, 77–78

technical images 67–69, 74, 77–79

technical language 15, 89, 118

technical meaning(s) 12, 48, 65–66, 68, 71, 75, 154–155

technical-superconglomerates 75

technical terms 47, 49, 51, 97, 148, 157, 209, 218

technical word(s) 66, 90, 94, 96–98, 153–158

Tett, L. 35

threshold concepts 103–122

trained gazes 9

transformative learning 119

transformative shifts 118

transition (from first to second year) 206–207, 209, 211, 216, 218, 220; to university 21, 35, 166

translation device 26, 46–47, 50, 66–67, 72, 79, 85–95, 97–99, 153, 156, 178–179, 212, 236, 238

unpacking 63, 90, 93, 95, 103, 163, 165–166, 215, 208, 212, 220

unpack/s/ed 5, 9, 22, 84, 90, 93, 97, 103, 165–166, 176, 209, 212, 215, 217

Valencia 158–159, 162

Virtanen, A. and Tynjala, P. 22

Walker, G. 106

'wicked problems' 129, 145

Whittaker, R. 35

Wilson-Strydom, M. 21–22

Wiggins, H.Z. 217

Wolff, K. 209

Wolfson, A.J. et al. 105

Wood, L. 218

Wood, L.N. 195, 217

wording tool 153, 155–156, 164

word-grouping tool 153, 156, 158–159, 164–165

worldly codes 10, 43, 54–55

writing proficiency 150

Zaky, H. 104–105

#FeesMustFall 233

For Product Safety Concerns and Information please contact our EU
representative GPSR@taylorandfrancis.com
Taylor & Francis Verlag GmbH, Kaufingerstraße 24, 80331 München, Germany